CHINESE MYTHS AND FANTASIES

Oxford Myths and Legends
in paperback

*

African Myths and Legends
Kathleen Arnott

Chinese Myths and Fantasies
Cyril Birch

English Fables and Fairy Stories
James Reeves

French Legends, Tales and Fairy Stories
Barbara Leonie Picard

Hungarian Folk-tales
Val Biro

Indian Tales and Legends
J E B Gray

Japanese Tales and Legends
Helen and William McAlpine

Russian Tales and Legends
Charles Downing

Scandinavian Legends and Folk-tales
Gwyn Jones

Scottish Folk-tales and Legends
Barbara Ker Wilson

West Indian Folk-tales
Philip Sherlock

The Iliad
Barbara Leonie Picard

The Odyssey
Barbara Leonie Picard

Gods and Men
John Bailey, Kenneth McLeish,
David Spearman

Chinese Myths and Fantasies

Retold by
CYRIL BIRCH

Illustrated by
JOAN KIDDELL-MONROE

OXFORD UNIVERSITY PRESS

OXFORD NEW YORK TORONTO

Oxford University Press, Walton Street, Oxford OX2 6DP

Oxford New York Toronto
Delhi Bombay Calcutta Madras Karachi
Petaling Jaya Singapore Hong Kong Tokyo
Nairobi Dar es Salaam Cape Town
Melbourne Auckland

and associated companies in
Berlin Ibadan

Oxford is a trade mark of Oxford University Press

First published 1961
Reprinted in 1962, 1969, 1971, 1975
First published in paperback 1992

With thanks to Sue Williams, and Mark and
Sylvia Bailey, for the Chinese embroideries
used on the cover of this edition.

A CIP catalogue record for this book is available
from the British Library

ISBN 0 19 274152 7

Printed in Great Britain

FOR

David and Cathy

Contents

THE CONQUERORS OF CHAOS

Heaven and Earth and Man 3
The Greatest Archer 9
The Quellers of the Flood 20

FAIRIES, GHOSTS AND OTHERS

The Dinner that Cooked Itself 37
How Marriages are Made 43
The Three Precious Packets 48
The Man Who Nearly Became Fishpaste 55
A Shiver of Ghosts 62
Magicians of the Way 80
The Inn of Donkeys 89
The Pavilion of Peril 95
The Rainmakers 101

THE REVOLT OF THE DEMONS

1 Aunt Piety 113
2 Eggborn 123
3 The Text 138
4 Eterna 152
5 Recruits 166
6 Revolt 176

Heaven and Earth and Man

EARTH with its mountains, rivers and seas, Sky with its sun, moon and stars: in the beginning all these were one, and the one was Chaos. Nothing had taken shape, all was a dark swirling confusion, over and under, round and round. For countless ages this was the way of the universe, unformed and unillumined, until from the midst of Chaos came P'an Ku. Slowly, slowly, he grew into being, feeding on the elements, eyes closed, sleeping a sleep of eighteen thousand years. At last the moment came when he woke from his sleeping. He opened his eyes: nothing could he see, nothing but darkness, nothing but confusion. In his anger he raised his great arm and struck out blindly in the face of the murk, and with one great crashing blow he scattered the elements of Chaos.

The swirling ceased, and in its place came a new kind of movement. No longer confined, all those things which were light in weight and pure in nature rose upwards; all those things which were heavy and gross sank down. With his one mighty blow P'an Ku had freed sky from earth.

Now P'an Ku stood with his feet on earth, and the sky rested on his head. So long as he stood between the two they could not come together again. And as he stood, the rising and the sinking went on. With each day that passed earth grew thicker by ten feet and the sky rose higher by ten feet, thrust ever farther from the earth by P'an Ku's body which daily grew in height by ten feet also. For eighteen thousand years more P'an Ku continued to grow until his own body was gigantic, and until earth was formed of massive thickness and the sky had risen far above. Thousands of miles tall he stood, a great pillar separating earth from sky so that the two might never again come together to dissolve once more into a single Chaos. Throughout long ages he stood, until the time when he could be sure that earth and sky were fixed and firm in their places.

When this time came P'an Ku, his task achieved, lay down on earth to rest, and resting died. And now he, who in his life had brought shape to the universe, by his death gave his body to make it rich and beautiful. He gave the breath from his body to form the winds and clouds, his voice to be the rolling thunder, his two eyes to be the sun and moon, the hairs of his head and beard to be the stars, the sweat of his brow to be the rain and dew. To the earth he gave his body for the mountains and his hands and feet for the two poles and the extremes of east and west. His blood flowed as the rivers of earth and his veins ran as the roads which cover the land. His flesh became the soil of the fields and the hairs of his body grew on as the flowers and trees. As for his bones and teeth, these sank deep below the surface of earth to enrich it as precious metals.

And so P'an Ku brought out of Chaos the heavens in all their glory and the earth with all its splendours.

But although the earth could now present its lovely landscapes, although beasts ran in its forests and fish swam in its rivers, still it seemed to lack something, something which would make it less empty and dull for the gods who came down from Heaven to roam over its surface. One day the goddess Nü-kua, whose body was that of a dragon but whose head was of human form, grew weary of the loneliness of earth. After long thought she stooped and took from the ground a lump of clay. From this she fashioned with her dragon claws a tiny creature. The head she shaped after the pattern of her own, but to the body she gave two arms and two legs. She set the little thing back on the ground: and the first human being came to life and danced and made sounds of joy to delight the eyes and ears of the goddess. Quickly she made many more of these charming humans, and felt lonely no longer as they danced together all about her.

Then, as she rested a while from her task and watched the sons and daughters of her own creation go off together across the earth, a new thought came to her. What would become of the world when all these humans she had made grew old and died? They were fine beings, well fitted to rule over the beasts of the earth; but they would not live for ever. To fill the earth with humans, then when these had gone to make more to take their place, this would mean an endless task for the goddess. And so to solve this problem Nü-kua brought together man and woman and taught them the ways of marriage. Now they could create for themselves their own sons and daughters, and these in turn could continue to people the earth throughout time.

After this gift of marriage from Nü-kua, further blessings came to man from her husband, the great god Fu-hsi. He again had a human head but the body of a dragon. He taught men how to weave ropes to make nets for fishing, and he made the lute from which men first drew music. His also was the priceless gift of fire. Men had seen and feared the fire which was struck from the forest trees by the passing of the Lord of the Thunderstorm. But Fu-hsi, who was the son of this same lord, taught men to drill wood against wood and make fire for their own use, for warmth and for cooking.

Already the creatures of Nü-kua's making could speak their thoughts to one another, but Fu-hsi now drew for them the eight precious symbols with which they could begin to make records for those who were to come after. He drew three strokes ☰ to represent Heaven; the three strokes broken ☷ represented earth. That symbol whose middle stroke was solid ☵ represented water, that whose middle stroke was broken ☲ represented fire. A solid stroke above ☶ gave the sign for mountains, a solid stroke beneath ☳ the sign for storm; a broken line below ☴ showed wind, a broken line on top ☱ showed marshland. With these eight powerful symbols man could begin to record all he observed of the world about him.

For long years men lived their lives in a world at peace. Then, suddenly, there spread from Heaven to earth a conflict which threatened to put an end to all creation. This was the battle between the Spirit of Water, Kung-kung, and the Spirit of Fire, Chu-jung. Down to earth came the turbulent, wilful Kung-kung to whip up huge waves on river and lake and lead his scaly hordes against his arch-enemy, Fire. Chu-jung fought back with tongues of flame and scorching breath and halted the rebel Water in his path. Kung-kung's armies dispersed and he, their leader, turned and fled. But his flight brought with it a peril greater yet. For, dashing blindly off to the west, Kung-kung struck his head against the mountain Pu-chou-shan, which was none other than the pillar that in the western corner held up the sky.

Kung-kung made good his escape, but he left the world in a disastrous state. Great holes appeared in the sky, whilst the earth tilted up in the west. In that region deep cracks and fissures appeared which are still to be seen to this day. All the rivers and lakes spilled out their waters, which ran off and still run eastwards: off to the south-east, where the earth had slipped down low, ran the waters together to form a vast ocean there. Meanwhile, out of the shaken mountain forests fire still raged forth, and wild beasts of every kind left their lairs to maraud through the world of helpless, terrified men.

It was left to the goddess Nü-kua to bring back order to the world, to quell the fire and flood and tame the wandering beasts. She it was also who selected from the beds of rivers stones of the most perfect colouring. These she heated until they could be moulded, then with these stones, block by block, she patched the holes in the sky. Lastly, she killed a giant turtle, and cut off its powerful legs to make pillars between which the sky is firmly held over the earth, never again to fall.

So the peace of the world was restored. But the mountains still rise in the west, and it is to there that the sun, moon and stars still run down the tilted sky; whilst to the east, the waters of the earth still gather into the restless ocean.

The Greatest Archer

THE greatest of all archers was Yi, for the targets of his deadly arrows were not made of straw, nor were they mere creatures of flesh and blood. The enemies Yi fought and conquered were powerful spirits who rebelled against the order of Nature, in the time of the saintly Emperor Yao. Under his rule men lived in peace, but these malevolent spirits took to themselves the forces of the elements and threatened to destroy all that lived.

It all began harmlessly enough, when ten children tired of playing each one by himself and decided that they would all go out to play together. For these were the children of the Supreme Ruler, born of his wife Hsi-ho. Each of these boys was a mighty star, a sun, and they lived all together above the Eastern Ocean. There a giant tree, the Fu-sang tree, rose thousands of feet from the surface of the water. A thousand men with arms outstretched could not span its trunk. It was in the branches of this tree that the ten suns took their rest. Each dawn it was from there that one of them would set out, as his turn came according to a fixed rota, on his journey across the heavens bringing light and warmth to earth. As they kept so obediently to their rota, men had never seen more than one sun at a time, and indeed they did not know that more than one sun existed, for all ten of them looked exactly alike.

No one knows quite why these ten sun-children suddenly took it into their heads to set out one morning all together across the sky. But the Supreme Ruler learned of this happening with alarm. He knew that his children brought great blessings to the earth—one at a time; but with ten in the sky together, surely only catastrophe could follow? His fears were confirmed when the Emperor Yao, in place of his customary prayers of thanksgiving, began to speak of blinding light, unbearable heat, of parched cattle and burning crops. All-powerful though he was on earth, even the Emperor Yao himself was helpless in the face of this new peril in the heavens. Assistance must be sent to him, some hero must descend who would save the world from the fate which threatened. At once the Supreme Ruler thought of his noblest warrior. Yi, the Heavenly Archer, had the skill, the courage and the goodness which the world would need. Yi it was, therefore, who on a night of full moon came down to earth and announced his presence at the gate of the simple dwelling from which Yao ruled his people.

Yao rose to welcome his visitor, and praised the great red bow he carried in his hand. He led him out into the street, where he pointed to the mountain peak which rose in the distance above the roofs of the capital. On its summit grew a solitary pine-tree. 'Let us see what use you make of this mighty bow you carry,' the Emperor commanded.

Slowly, deliberately, the great archer selected from

the quiver at his side an arrow, long and straight and tipped with purest, hardest bronze. This he fitted to the string, which he drew back with one smooth gliding of his right arm. Legs apart, body upright, he faced the distant peak. The pine gleamed in the moonlight, a thread of silver. Yi took his aim. There was a sound like the plucking of a giant zither as he released the string of his bow—and the pine on the hill-top clove in two before the impact of the speeding shaft.

The Emperor smiled. 'Take your rest now,' he ordered. 'Tomorrow there is much for you to do.'

Yi spent the night on a bed of fragrant grasses. When the Emperor himself came to wake him the sky was not yet light. Yi knew that Yao had not slept: he must have spent the night in prayer for his people's deliverance. Still there was no sign of weariness in the compassionate lines of his face. 'Come now,' said the Emperor calmly. 'I wish you to witness the dawn.'

Unattended by any suite the two walked out through the streets of the sleeping capital. The watchman at the gate, wary at first of strangers at such an hour, threw himself flat on the ground when he recognized his Emperor. Yao raised him to his feet, and bade him light their way with his torch to the top of the gate-tower. No sooner had they reached this vantage-point than the first washes of silver on the eastern horizon began to seep into the deep blue of the night sky. The dawn was breaking. For a few moments all was just as in any other dawn. But soon it became apparent that a greater light than the light of the morning was waiting below the horizon. The silver washes were now a flood, and the silver itself was of heightened brilliance. In a second, it seemed, the sky gleamed gold in the east. At the centre of the gold, the disk of the sun lifted itself clear of the land-line. Then, at the precise moment when the sun stood clear in the sky, over the horizon flashed the rims of two suns, each of equal brilliance with the first. Now about this group of suns the sky flamed angrily for a while, until yet more suns rose, more than the watchers were able to make out in the liquid, boiling, white glare that filled the heavens.

And it was hot. Yi was astonished, when he put his hand on the stone rampart, to find that it burned and blistered, already in the early dawn. He opened his mouth to speak, but the parched air dried his mouth at once so that only a croaking sound came out. The Emperor Yao understood, and nodded and signed that they were to go down. But before they left the roof of the tower, he pointed to a field of millet below, close against the city wall. The ears had not yet formed on the stalks, and the stalks should have been green. But they were brown, withered. As they watched, from a corner of the field came wisps of smoke, and in a few moments the field was aflame. The flames themselves lasted only for a minute. Then all that was left was a layer of white ash, beneath which the ground was already beginning to crack.

Quickly the two men descended from the roof. As they left the tower they felt the soles of their feet burn against the ground. Yi turned back to the tower, where in the shade he found a pail which still contained some water. With this they soaked their sandals, and thus prepared they were able to make their way, though with pain, back to the Emperor's dwelling.

There Yao told his visitor of messengers who had come scores and hundreds of miles from all corners of the empire. They had brought stories of whole villages dying of thirst when the wells had dried; of men perishing in the fields from exhaustion in the heat; of forests ablaze, dense clouds of steam over boiling lakes, and even, from the south, of valleys overwhelmed with lava from the melting mountain-sides.

News such as this was terrible to hear: but worse was to come. While the Emperor was still speaking there was a commotion outside the room and a man burst in. Another messenger—like the others, pale with fear beneath the stains of travel. His tale was the worst yet. Out of the unnatural happenings had come an unnatural creature, a monster with the giant body of a man and the head of an animal. It had one vicious weapon which none could resist: a great tooth, fully six feet in length, as sharp at the cutting edge as the edge of a chisel. This monster was creating havoc in the south, tearing down the people's flimsy huts before it tore at their bodies.

Hardly had the man finished his report when another came. 'Your Majesty, give ear to the sufferings of your people,' he cried. 'In the east a great bird flies, a peacock with mighty wings whose beating raises terrible raging winds. Trees are torn out by the roots, houses swept away like swirling autumn leaves, and the people cower, wretched and afraid, in caves and holes in the ground.'

And still a third, again exhausted by his journey through the fierce heat from the ten suns. 'The wide Tung-t'ing Lake,' reported this man, 'is terrorized by a monster, a sea-serpent whose passage through the water brings storm and flood. No one knows how many fishermen of the lake have disappeared down its cavernous throat nor how many whole villages have been swamped in its wake.'

Yi looked at the Emperor's expression and knew that he was suffering not for himself alone but for all his people. 'There is only one way,' the great archer said. 'First, the false suns must perish, or all life is at an end.'

Then Yi set to work. Carefully he chose ten arrows, which he placed in his quiver. He took up his bow and tested it, then strode out into the blinding light, into the choking heat. The Emperor followed him, and when they reached the market-place they found the people of the capital already assembled there, waiting patiently for the saviour of whom the watchman of the city gate had told them.

Yi closed his eyes for some seconds against the searing sunlight. Then he placed an arrow against the bow-string and drew to the full, his mighty shoulders forced back and the muscles leaping on his arms. Squinting upwards he loosed his shaft. There was a breathless pause: then, across the white pool of light of the sky leapt tongues of red. Now above the heads of the watchers floated tiny specks—the feathers of a bird; another instant, and something huge and black plummeted to the ground at the feet of the great archer. It was a monstrous crow, and it was transfixed by an arrow. This, then, was the spirit of the sun, a great crow just as the old magicians had always said. 'The golden crow'—that was what they had called the sun: but now the crow was dead, and black, pierced by the arrow of Yi.

Still the sky blazed and the earth burned. Nine suns remained, nine suns which at any moment might transform themselves into giant crows and take wing over the horizon, beyond the reach of his swiftest arrow. With unhurried movements, but losing not a second, he loaded, drew and released, loaded, drew and released. Two more explosions in the sky, soundless at that great distance; two more flurries of feathers, two more black bodies crashing to the ground. On and on laboured Yi, muscles bulging, shoulders aching from the prodigious strain of his bow. The air was filled with

the singing of his bow-string. Now four suns remained, and could be counted clearly against a sky less molten in its brilliance. The burning had almost ceased. The faces of the watchers were no longer seared blind with pain, but were lit with half-unbelieving hope.

The eyes of the Emperor Yao fell on Yi's quiver. Four arrows still remained—and Yi was conscious of nothing but the need to shoot the suns from the sky. But one sun must be spared, thought Yao, if the world were not to be plunged into eternal night and winter. Quietly he withdrew one arrow from the archer's quiver, and placed it in his sleeve. The seventh sun fell from the sky, the eighth, the ninth—and Yi's hand reached to his quiver and found it empty. Sweating, weary, he looked at the Emperor Yao, then up to the clear blue sky, to the one welcome sun which has given its blessings to the world through all the ages since that moment. Yi looked at the sun; and then his ears were filled with the shouts of the crowd of watchers, and he smiled and laid down his bow.

Already, at the edges of the sky, white clouds were forming, clouds of happy omen which would soon bring rain to the parching earth. Yi, after his heroic labours, would dearly have loved to rest. It was pleasant here, tempting, under the now kindly sky. All the world was his friend, for he had saved all the world from destruction; and the saintly Emperor Yao was even now giving orders for the preparation of a feast for the hero. But Yi well knew that his labours were not yet ended. In the south and in the east still roamed the monsters, Chiseltooth, the Windbird, the serpent of the Tung-t'ing Lake. Pausing only to replenish his quiver and gird on his sword, Yi called for the messenger from the south and started in pursuit of Chiseltooth.

It was a long, hard journey. Across the plains strode the great archer and his companion, over the passes, through the swamps and forests. Boatmen ferried them across broad rivers, guides led them to the fords of a hundred streams, until at last they reached the mountains of the south where Chiseltooth had his home. Here they began to find the signs of the monster's presence. On the ground lay bodies from which the heads had been severed by that razor-sharp fang; huts near by had been ripped from their foundations. Even as he looked, Yi heard behind him an angry roar. He turned —and there stood his quarry, the one huge tooth gleaming yellow against its chest.

In a flash Yi had raised his bow, an arrow fitted to the string. But before he could draw back the string Chiseltooth had disappeared into a cave behind him. Warily Yi approached the cave. The monster re-emerged from its entrance, and this time little of him could be seen behind a massive shield. Yi stood his ground and waited, his bow still raised, arrow poised on the string. Chiseltooth lumbered nearer, very slowly, nearer, until it seemed the two must collide. Then, with a swift movement, he lowered his shield and the great fang reared up to strike. In that second Yi loosed the arrow he had held poised for so long. It flew straight to its mark at the root of the monster's tooth. At such a range the impact was like that of a thunderbolt. The tooth snapped off, and the monster, making not a sound, fell to the earth, dead.

The sun was sinking in the west, but Yi was not yet ready to take his rest. In the cool of evening he turned his back on the sunset and set out in search of the Windbird. After much journeying he came to a wide river, and down this he sailed, ever eastward. At last he met with the signs he had been looking for, signs of destruction by raging winds, crops flattened, trees and huts and boats all tossed about like straws. And then he heard a far-distant rustling, and peering up made out, far off in the sky, a speck of black—the Windbird was coming. He paddled in to the bank, sprang ashore and hid himself in a thicket which still stood and would be in the path of the approaching bird. As he waited he reasoned with himself, 'One arrow may not kill a bird of such a size. And if it is merely wounded, and flies off to its nest to wait till the wound is healed—it may do great harm thereafter, before I find it again. The answer is to capture it now, and make an end of it now.'

To do this he used a method which hunters of birds have followed ever since. To his straightest arrow he attached a long, strong cord, the other end of which he tied to his bow. Then he fitted arrow to bow-string and waited in the shelter of the thicket. Very soon the rustling he had heard had grown into a roaring, and the roaring into a deafening thunder, and now the sky was darkened and the giant peacock was overhead. The great archer shot, the arrow struck home, and the bird leapt upward in its flight. Still it flew on, and Yi braced his legs against the earth to withstand the storm from its beating pinions, while all about him tall trees shook and toppled before the howling wind. The tug as the line tautened on the bow sent a searing pain across his shoulders, but still he held on; until at last the Windbird's effort failed and it dropped to earth, captive at the end of Yi's cord. Yi ran to it, and as it lay thrashing its wings against the ground he drew his sword and cut the peacock's head from its body.

The Windbird was dead, and Chiseltooth was dead. But on the Tung-t'ing Lake the sea-serpent still reigned in terror. Once more Yi took up his bow and set out on his way. After many days he reached the shore of the lake, which is less a lake than an inland sea. There he took a small fishing-boat, and launched out alone on the face of the waters. Out into the lake he sailed, and hard and long he searched until he saw across the water before him coil after coil of the scaly monster rising and falling in awful succession. Closer he sailed, while the waters of the lake grew ever more turbulent. He took up his red bow. His first arrow pierced the neck of the serpent, but the only effect was that the waves rose higher than before. Again and again he shot, until the serpent's body bristled with arrows and it writhed in anguish. Yi's little craft tossed and plunged on billows high as hills, and now it was impossible for him to shoot again. He drew his sword as his boat swept up a towering crest of water. Then, as it trembled on the brink of a sea-green precipice, he leapt, down, down, on to the broad and slippery back of the monster.

Then followed the most desperate contest Yi had known. Time and again his sword plunged to the hilt in the serpent's body; time and again he kicked his own body out of the way of the menacing fangs which reared above him. The bubbling, surging water of the lake was no longer green, but stained dark red with the sea-serpent's blood. The end, at last, came suddenly: one final lunge of the hero's sword struck deep into the monster's vitals. The wicked, scaly body went limp, then quivered, and sank at last beneath the crimson waves.

Wearily Yi allowed himself to float to the lake-shore. His task was ended. Nine false suns had perished. Three monsters such as the world had never before seen had been destroyed. The saintly Emperor Yao could rule again a world of men at peace. Yi, the great archer, could take his rest.

The Quellers of the Flood

LIKE the lands of the Bible in the time of Noah, China in ancient times knew the wrath of Heaven in the form of a terrible, all-consuming flood. For long the Yellow Emperor, Ruler of Heaven, had been caused sorrow and anger by the wicked ways of men on earth. At last he came to feel that only the severest punishment would succeed in bringing the mortal world back to its senses, and as the instrument of his punishment he chose to send down torrential, endless flood rains. The cruel and vindictive Spirit of Water, Kung-kung, was placed in charge. Kung-kung carried out his duties without mercy.

First the rain pinned the people in their houses, making it impossible to go out into the fields. Paths turned into quagmires, pools formed in every hollow. Soon, whatever was not under shelter was soaked through and rotted: stacks of grain, piles of fodder, all were useless after a few days of the rain. Holes appeared in the flimsy thatches of huts, and the people shivered and moaned with cold and hunger.

Then, one after another, the rivers broke their banks.

The dwellers in the plains and valleys, and all who lived by lake or sea-shore, saw the waters surge towards them or slowly rise at their feet. Many went no farther than the roof of their hut before the waters overtook them. The rest made for the hills, there to seek shelter in caves dug out of the wind-blown soil. Some even copied the birds and made rough nests for themselves in the topmost branches of trees—anything, anywhere to reach a height, to be out of reach of the swirling, menacing waters. Carts and chariots no longer had any value. Everyone wanted a boat, and the boat-builders worked day and night. Every man became a fisherman, for meat was no longer to be had. Every woman searched all day for a tree with leaves or bark on it that her family could eat. And all there was to drink, now that the wells and streams had disappeared, was the brackish, muddy water of the flood itself.

Terrible indeed was the wrath of the Yellow Emperor. But there was one spirit who looked down on the world of men and was moved to pity by their plight. This was Kun, the grandson of the Yellow Emperor, a spirit known to men in the form of a white horse. Kun could not bear to stand idly by and watch such suffering. But when he pleaded with the Ruler of Heaven, his grandfather, to withdraw the flood of his anger, his plea was refused: the measure was not yet full. Kun left the imperial presence and stood wrapped in distress and perplexity.

As he stood there, at a loss, he saw two curious creatures making their way slowly towards him. One was a horned owl, the other was a black tortoise, and they were helping one another along.

'What problem is causing you such distress, great spirit?' asked the owl as the two approached.

'I am filled with pity for the suffering humans,'

replied Kun, 'and yet I do not know how their world is to be freed from the great flood.'

'Easy,' grunted the black tortoise. 'All you need is the Magic Mould.'

'The magic mould!' exclaimed Kun, astonished. 'What sort of a thing is that?'

'It's what I said, it's Magic Mould,' answered the tortoise.

'That's right,' added the owl, helpfully. 'It's what he said. It's mould, you see—earth, soil, you know; and it's magic. That means it can grow, by itself, to any size you want. You only need a little bit. Just get hold of a piece and see for yourself.'

'But where do you get it from?'

'Your grandfather,' said the owl. 'He keeps it.'

Kun's hopes, which had been raised so high, were dashed immediately. 'He would never let me have it,' he sighed.

'Easy,' grunted the black tortoise again. 'Steal it.'

Kun started in alarm, but the owl motioned to him to say nothing, then began to whisper in his ear. The tortoise, too, added a word now and again, and Kun began to nod in agreement with what they were saying. Soon, between the three of them, they had worked out a plan for stealing the Magic Mould from the Yellow Emperor.

Now unfortunately none of the old books tells us just what the plan was; they only tell us that the plan succeeded, so that Kun did in fact lay his hands on a good supply of the precious earth. Although we can't be sure exactly how he did it, it is clear that a spirit has many advantages when it comes to breaking and entering. However deep in the palace the Magic Mould was hidden, however strong the lock and thick the wall and fierce the guards protecting it, we must remember that

Kun had only to change himself into a puff of smoke to get past the guards. Having got past the guards he had only to change himself into a badger to tunnel through the wall; and having tunnelled through the wall he had only to change himself into a tongue of fire to melt the lock on the box which contained the treasure he sought.

As soon as the Magic Mould was in his possession, Kun descended to earth and began his fight against the flood. From a mountain peak he looked down on a broad sheet of water which concealed what had once been a broad and fertile valley. Kun broke off a tiny lump, not much bigger than a pea, from the precious soil he carried, and threw it into the water, ordering it to grow until he told it to stop. The little crumb disappeared beneath the surface. Kun watched. Soon a shadow formed beneath the water; in a little while, it became apparent that the lake bottom was rising. What had been a fathomless lake was now merely a shallow sheet of water. At point after point, soil broke through to the surface—the lake was a marsh, Kun was looking down on a valley floor that was no more than waterlogged. Still the Magic Mould grew and grew until all the water was absorbed; and when at last the valley was filled with brown, rich, dry soil, Kun ordered it to stop. Delighted he surveyed his handiwork. Then, from caves on the mountain-sides, poor wretches of human beings came out. Dazed, unbelieving, they looked down on the valley. They had scarcely the strength to cheer, but one after another they stumbled down the hill to touch the good brown earth so magically granted them. And some turned back to their caves to bring out bags of precious seeds which they had guarded through all their vicissitudes, and which now at last they could sow.

After this first experiment Kun travelled, unwearying, through the length and breadth of the land. He filled up the valleys, he raised the plains; wherever he cast his Magic Mould the flood waters were soaked up as though by some giant sponge. Across the flat plains he built dykes by ordering the Magic Mould to grow out lengthwise; then, behind the dykes he would create new fields and pastures for men to work.

But the wrath of the Yellow Emperor was not to be so easily set aside. When it was reported to him that the Magic Mould had been stolen his anger was renewed. This was rebellion in Heaven, treachery by his own grandson! There could be only one punishment, the punishment of death. The Yellow Emperor summoned the Spirit of Fire, Chu-jung, and gave him a dreadful commission. Chu-jung in his turn was to go down to earth, and there he was to execute the Emperor's grandson Kun and bring back the Magic Mould which he had stolen.

The warlike Chu-jung lost no time in carrying out his commission. Kun was no match for him. He pursued him across the great spaces of the earth, and ran him down at last on Feather Mountain, which is near the North Pole. Here all is cold and gloom, for the light of the sun never penetrates these wilds. The darkness is relieved only by the faint glimmer of a candle, held in the mouth of the Candle Dragon who stands guard over the polar regions. Not far from Feather Mountain is the City of Silence, the resting-place of human souls after death. In this land of ice and darkness Chu-jung killed Kun, and left his body where it lay on Feather Mountain.

But it is not such a simple matter to put an end to a heavenly spirit. The strength of Kun's compassion and the strength of his ambition had been sufficient to drive him over the surface of the earth reclaiming land from the flood. Although his physical body now lay lifeless on the slope of the hill, this compassion and this ambition lived on. By their power the dead body was protected from decay, so that after three years had passed no change was visible. Yet a change had taken place: for within the body of Kun, as it lay in the icy waste, the great will of the murdered spirit was creating a new being—a son! Gradually, under the surface, all the strength

which had belonged to Kun was going into the making of a new creature. For still the flood rains fell. Despite all the labours of Kun the floods were not yet conquered. In many places the people still cowered in cold and hunger and misery on the bleak hill-sides. And even some of the new pastures Kun had created were slowly sinking again under the face of the water. The work that Kun had started must be continued.

It must have been something of this kind that the Yellow Emperor feared. For when it was reported to him that the body of Kun had lain on Feather Mountain for three years and still had the appearance that belonged to it in life, he knew that there was danger in this. He feared that the rebel against his authority, the traitor within his own family, though executed by Chu-jung, might still be at work against him. Therefore he ordered a second spirit to descend to the polar region, there to destroy the body of Kun. To accomplish this task the Ruler of Heaven furnished the spirit with the famous Sword of Wu, which nothing could resist.

Down went the spirit, flash went the Sword of Wu: and thus came to the light of day the being which had been nurtured in silence and secrecy for three years inside the corpse of Kun. As the sword struck, cutting open Kun's body, out sprang a horned dragon, young and strong, which spread its powerful wings and soared up into the dark sky. This was the son that had been born of Kun's will to save mankind from the flood. This was Yü the Great, who was destined finally to bring the waters under control.

Straight to the courts of Heaven flew the dragon, on and on until at last he reached the presence of the Yellow Emperor himself. There, Yü made obeisance, and in all humility sought an audience with his mighty ancestor. This the Yellow Emperor granted, and waited on his throne for Yü to begin.

'I entreat Your Majesty to consider,' said Yü, 'whether the punishment suffered by the mortal world has not already been sufficient? For long ages now, those who did not perish have lingered on, starving and destitute. Surely the wickedness men practised has now been driven out of them? The work of my father——'

'You had better not speak of your father,' the Emperor broke in. 'He was a rebel against my authority. My own grandson, he stole from me the Magic Mould and set my works aside.'

'He stole the Magic Mould, and in that he was at fault,' said Yü. 'But I come to beg you to give me the Magic Mould, from your own hand, so that I may put an end to the havoc on earth and save the world which our ancestors created from utter destruction.

In your boundless wisdom and compassion, hear my request.'

The Yellow Emperor reflected. He had in fact been wondering for some time past whether Kung-kung, the spirit he had appointed long ago to take charge of the floods, had not been too severe in the tribulations he had inflicted on humanity. Kun's rebellion had angered him; but this youth approached him in the proper attitude of submissiveness. He inclined to grant the request. And indeed, if he did not grant it, who could tell what new rebellion might not follow? All the hot strength of Kun had passed into the young Yü in the three years of his creation. The peace of the heavenly family was in danger if such a spirit should set himself against his ruler.

'Your request is granted,' pronounced the Yellow Emperor. He beckoned, and a black tortoise slowly made its way towards the throne—the same black tortoise who long ago had advised the spirit Kun, and who had since received forgiveness. 'You may have as much of the Magic Mould as can be piled on the broad back of this creature, and he shall carry it for you about the earth.' The Emperor beckoned a second time, and a winged dragon entered and bowed before him. 'Here is the slayer of the rebel Ch'ih-yu. He also shall assist you in your work. As for your father: his strength has now passed into you, and his task will be accomplished by you. He is granted the form of a yellow dragon, and he is to retire into the Feather Grotto for his abode. Now go, and do not fail in your labours.'

Joyfully Yü left the imperial presence, collected as much of the Magic Mould as the black tortoise could carry, and then in human form returned with his companions to earth. But on his arrival there he found a turmoil and a chaos worse than was ever known before.

Kung-kung, the Spirit of the Waters, was angered by this new threat to his execution of the Yellow Emperor's sentence. He had shown his anger by shaking and stirring the waters which covered the earth. What had been endless placid lakes were now contorted into raging seas. The tiny boats in which people still sought to ride the floods were overwhelmed by the towering waves. Tall trees, in whose branches men nested like birds, were lashed by the flying spray. The turbulent waters lapped into hill-caves which had formerly been untouched. This was the worst trial which had yet faced mankind, and the dreadful noise of wailing could be heard above the raging of the storm.

Yü saw that before he could set to work he must first vanquish Kung-kung. So he took up his post on the mountain Hui-chi, and there he summoned all the spirits to present themselves before him. The air was filled with a rushing sound as the spirits assembled to hear Yü's bidding. All came with the exception of Kung-kung himself and one other—the Stormguard. The Stormguard was late, and Yü suspected him of treachery. He acted at once, for his task was too stern to permit the taking of risks. He executed the Stormguard on Mount Hui-chi, as a dreadful warning to the others. They buried his body; many centuries later, a giant bone was dug up on Mount Hui-chi, and this was recognized by the people of the time to be a bone of the Stormguard. The single bone could be lifted only with difficulty by many men, and when it was moved it made a heavy load for a large cart.

Yü now ordered the assembled spirits to attack the Spirit of the Waters. Kung-kung was filled with terror when he saw the great army approaching, and fled—and no one has ever learnt his hiding-place. Now at last Yü the Great could resume the work which his father

Kun had been forced to leave unfinished. Followed by the black tortoise, the bearer of the Magic Mould, he moved about the earth. Like his father before him he filled the valleys, built dykes across the plains, and behind the dykes created new lands for the farmers. Where the waters were shallow he raised mountains, which are the peaks and ranges we know today. All the time Yü drove before him the Winged Dragon. Where the Dragon's tail touched the ground there was drawn the line of a watercourse, and thus Yü fashioned the great rivers to draw off the flood waters from the plains.

For thirty years Yü laboured, never ceasing. But at last he began to reflect on his own life, and on his need of a wife to share it.

'Surely,' he said to himself, 'there exists somewhere in the world the woman who is destined to be my wife, the partner of my days on earth. But how shall I recognize her? Will there not be some sign to lead me to her presence?'

As these thoughts were passing through his mind he was approaching a copse of willows. Suddenly from between the trees a small creature ran out towards him. It was a fox, a white fox, with nine tails. As it turned and ran off Yü recalled a song which was sung by the people of the district. The words were curious, and formed a prophecy:

The royal throne will ensue to him
Who sees the white fox with nine tails.
A flourishing line will ensue to him
Who marries the girl of T'u-shan.

The white fox surely was the sign. Although he did not fully understand the prophecy, Yü made his way straight to the hill T'u-shan, and there, sure enough, he found a maiden, gentle and fair and with wisdom in her eyes. This maiden was Nü-chiao, and Yü married her on the mountain Hui-chi, where he had first begun his labours. But not for Yü was the peaceful life of the family. He returned at once to his task, still uncompleted. Nü-chiao accompanied him on his travels through the world, and assisted him in his work.

The birth of Ch'i, the son of Yü and Nü-chiao, took place in a strange manner. When Yü surveyed the floods about the mountain Huan-yüan, he made the decision that the mountain must be tunnelled through. It was a formidable mass of rock, and in order to dig the tunnel Yü decided to transform himself into a giant bear. With strong sharp claws he tore at the rock, and soon was so absorbed in his task that he quite forgot how Nü-chiao had promised to bring him refreshment in the afternoon. Deep into the core of the mountain he tunnelled, and the sound of Nü-chiao's voice surprised him, echoing down the tunnel from where she stood far away on the open hill-side. He held his huge paws still and raised his shaggy head to listen: yes, it was his wife. Delighted, he bounded back along the tunnel he had dug. But in his excitement he never thought to resume his normal form. Nü-chiao, waiting at the entrance, was suddenly confronted not with her husband, but with a huge ponderous bear bursting from the bowels of the hill.

Terrified, she turned and ran. Yü called to reassure her—and with horror she realized that this was her husband. Perhaps this was his true form, which up to now he had disguised from her! In her heart feelings of shame mingled with the fear which possessed her. Faster and faster she ran, but always the heavy padding of the bear beat the earth close behind her, and drew ever nearer. In this way, Nü-chiao fleeing, Yü in his bear's form pursuing, they covered many miles. The woman's heart felt near to bursting, and her strength began to fail. She stumbled, picked herself up and ran on, many times. And now she saw before her a steep slope rising. It was the Lofty Peak, and she knew she could never ascend such a height. Desperate, she sank to the ground at the foot of the Lofty Peak, and there transformed herself into a boulder which would be safe from attack by her husband, the bear.

All this time Yü had been sorely perplexed by his wife's flight. Now, as he came to a standstill before the boulder, only one thought came to him. He reared up on his hind legs and bellowed, 'Give me my son!'—in a voice which shook the heavens.

In response to his cry, the boulder split open. And that was the manner of the birth of Ch'i. Ch'i, the son of Yü, born of the splitting of a boulder, was a handsome youth who wore jade ornaments and rode two dragons. He was a lover of music, and it is to him that the world of men owes its rarest harmonies. For Ch'i often rode his dragons up to the courts of Heaven. There, at feasts, he heard such music as was not known to men on earth. This music he noted down, and secreting it on his person brought it back to give to the world.

But this belonged to a later time. Yü had not yet accomplished his labours. When he had raised the mountains, etched out the rivers and banished the floods from the fertile plains, he undertook to measure the expanse of the earth. To do this he commissioned two spirits, one to pace the earth from north to south, the other to pace the earth from east to west. When the two spirits returned to make their report, it was discovered that each had covered two hundred and thirty-three thousand five hundred *li*—three *li* make one mile—and seventy-five paces. The earth, therefore, had the shape of a perfect square. During their survey the spirits had come across many bolt-holes of the flood waters, crevices and fissures into which they had retreated to await their chance to surge forth again. The number of bolt-holes larger than eight paces across was two hundred and thirty-three thousand five hundred and fifty-nine; all of these Yü stopped up, but the smaller ones were so great in number that he was unable to deal with them all. That is why the world is still plagued by floods, but on a scale which is minute in comparison with the great catastrophe of the time of Yü.

Wishing to provide a residence on earth for the gods on their visits, Yü dug up the K'un-lun Mountains from Heaven and transported them to earth. He set them down on the western frontier of China, and there they remain to this day. They form a ladder from earth to Heaven. For of the three peaks, if a man climb the first he will attain immortality. If he climb the second he will win magic powers and be given command of the winds and the rain. And if he climb the third he will reach Heaven and become a spirit.

The heroic achievements of Yü the Great were hailed with gratitude by the people of the world. The prophecy of the old song was fulfilled. They made him Emperor, and the dynasty thus founded, the Hsia, reigned for no less than four hundred and thirty-nine years. Yü himself reigned for only eight of these, for he was already an old man when he came to the throne. He died at last on a visit to the mountain Hui-chi, where he had married Nü-chiao. There his body was buried; but his spirit returned to Heaven to live for ever.

The great cave where Yü was buried can still be seen on Hui-chi. And there are other traces left to us of Yü's heroic work. On the border between the provinces of Shansi and Shensi, cliffs like beetling eyebrows overhang each side of the river. Below these jutting cliffs is the hollow through which Yü drained the flood waters from the region, in those years so long ago. The place is known by the name of Lung-men, the Dragons' Gate. For at a certain time each year, all the fish of river and sea assemble at this point. They hold a contest, each attempting to leap from the water clear across the overhanging cliffs. Those fish whose effort fails fall back into the river, where they remain. But those which clear the Dragons' Gate are transformed at once into dragons and continue their leap, soaring up into the blue vault of the sky.

FAIRIES, GHOSTS AND OTHERS

The Dinner that Cooked Itself

Of all the quests a young man must follow, none is more difficult than the quest for the girl who will marry him, be his wife and live with him happily ever after. And in China in the old times, he would have many an obstacle to overcome before his search ended. So it was that Hsieh Tuan failed time and again to find himself a bride.

Tuan was a straight, handsome youth of eighteen. He was only a humble clerk in the magistrate's court, but he worked always with a will, and his honesty and respectful bearing impressed all who knew him. His father and mother had both died when he was very small, but a kind neighbour, old Wang, had taken him in and treated him as one of his own sons. Now that Tuan had reached manhood he moved into a little house of his own. Not far from the house he had a small strip of land, where he tended his rice-plants and beans each day when his duties at court were done.

It was old Wang who hired a go-between to seek out a wife for Tuan. The go-between suggested Miss Ch'en, the pretty daughter of a farmer on the outskirts of the town. But when the dates of birth were compared, it was found that Miss Ch'en had been born in the year of the cat. Now Tuan belonged to the year of the dog; and with cat and dog under one roof there would never be a day's peace in the house. The go-between tried again a little farther away. But what was wrong this time was the characters with which the young lady wrote her name. They contained a sign which meant 'wood', and Tuan's name contained 'earth'. Now wood overcomes earth as a wooden plough turns a furrow, so that Tuan married to this girl would never have been master in his own house.

One match after another was considered and rejected. Even when everything else was right, people would object that Tuan was too poor to marry their daughter. But none of this made any difference to Tuan's daily work, which began at cock-crow when he left for the magistrate's court, and ended at nightfall when he returned from cultivating his tiny field.

One night Tuan, his hoe over his shoulder, was making his way slowly along the narrow path which joined his field to his house. He had worked even later than usual, and at dark sat down on a dyke and waited for the moon to rise. The moon was full, and its light now lay like snow on the thatch of his own cottage and sparkled like frost on the tiled roofs of his wealthier neighbours. He looked down to follow the winding of the narrow path. The moonlight glimmered on a stone by the edge of the path at his feet. But the stone had never been there before. And was it really a stone after all? It was rounded, and pointed at the top. He bent down for a closer look. Not a stone, but a snail—an enormous, giant snail, quite the size of a small bucket!

Of course, it was a sign of great good luck to find such

a rarity. Delighted, Tuan raised the snail gently in his hands and hurried on home. On the way he picked some succulent leaves for it to eat, and these he put together with the snail in a large, roomy earthenware storage-jar which stood just inside his door. He went to bed still rejoicing over his good fortune. In the morning when he woke, his first thought was to look inside the jar, and he was pleased to see that the snail had eaten a hearty breakfast.

And now a very curious thing happened. Tuan went off to the court as usual and came back home in the afternoon to have a bite to eat before going out to his field. But when he entered his little house he found the table set with bowl and chopsticks. Steam rose temptingly from a dish of cooked rice and vegetables, and on the newly swept floor was his large washing-bowl, filled with hot water ready for him to use!

'How kind people are,' Tuan thought as he washed his hands and face and squatted down to attack his dinner. 'It must be Mrs Wang who has stolen in here secretly to give me such a pleasant surprise. What a thoughtful thing to do!'

Never in his life had he tasted such delicious cooking. As soon as he had washed up he hurried round to Mrs Wang's house to thank her. But he was mistaken. Mrs Wang had not been near his cottage, and could not imagine who it might have been.

Yet when he returned from the court again the following day, exactly the same thing had happened again. Tuan was greeted as he entered the house by an even more appealing smell from the table, where a dish of little fried balls of pork lay beside the steaming rice. While he hungrily plied his chopsticks, Tuan tried hard to decide who it might be. At last he hit on an idea—it was the pretty Miss Ch'en, who could not be his wife because she was born in the wrong year, but had taken pity on his loneliness and decided to help him until he should have a wife of his own.

So round he went to the farm on the outskirts of the town, to thank Miss Ch'en for her kindness. But Miss Ch'en seemed to know no more about it than Mrs Wang had done. When he pleaded with her to tell him the truth, she began to tease him, 'I don't believe your story at all. I think you really have a wife already but are keeping her secret. She is in your house all the time, but you won't let anyone see her and are pretending it is someone else who is cooking your meals for you!'

Tuan returned home more mystified than ever. On his way he picked some fresh shoots of bamboo to feed his snail, which still seemed very happy to roam about inside its huge earthenware jar.

Every day for over a week the same thing went on happening. Each day Tuan would come home to find his room swept, his washing-water all heated and ready and his dinner waiting on the table. The news spread among all the neighbours, but no one offered to explain the mystery. At last Tuan determined to get to the bottom of it. He left his house as usual at the first crow of the cock. But as soon as the sun was up he came secretly back, and hid outside the fence to watch what might happen in the house.

For a while all was still. Then, suddenly, there was a movement: through the doorway he saw a hand appear out of the huge earthenware jar. After the hand another hand; and a lovely young girl, beautifully dressed in a silk robe, climbed out of the jar and crossed the room to the stove in the corner.

Quickly Tuan left his hiding-place and entered the house. His first concern was to look inside the storage-jar—no snail was there, but only an empty shell! In the corner by the stove the girl pressed herself against the wall in alarm.

'Who are you?' asked Tuan. 'Where do you come from, and why are you looking after my house for me?'

The girl said nothing. Her frightened glance darted about the room, and she tried to dash across to the earthenware jar, but Tuan prevented her.

Finally she spoke. Her voice was clear and sweet like a tinkling of jades. 'I am a fairy,' she said, 'and my

name is White Wave. The Lord of Heaven took pity on you because you are an orphan and live alone. And because you work hard and are honest and polite, he sent me to look after you. I was to stay with you for ten years, until you grew rich and married a wife. Then I was to leave you and return to Fairyland. But now you have spied on me in secret, and you have seen my true form. This is not permitted to a mortal. I must leave you at once. You must continue to work hard at court and cultivate your land with all your strength. But you may keep the shell which I left in the jar. Use it for storing rice, and empty it only when hunger threatens. You will find that it will at once fill up again.'

In vain Tuan pleaded with her to stay. No effort would avail, she had to go. The sky darkened and a storm blew up, the wind howled and rain lashed the roof. White Wave ran lightly across the room and out of the door, spread her arms wide and soared away, borne by the raging wind. As suddenly as it had started the storm came to an end, and Tuan was left in the calm morning, his eyes filled with tears.

Hsieh Tuan built a little shrine to the fairy White Wave, and he did not fail to sacrifice there on feast days. From that time onwards he was never short of food. And although he never became outstandingly wealthy, he married a wife at last who made him very happy all his life. Nor did he always remain a humble clerk, but ended up by becoming a district magistrate himself.

How Marriages are Made

ANOTHER seeker after marriage was Wei Ku. His parents died when he was still a boy. As he grew to manhood he longed to marry a wife so that he could have a real home. He searched far and wide for a bride, but it was no simple matter and something always seemed to go wrong. At last a friend to whom he had poured out his troubles promised to act as go-between to arrange his marriage with the daughter of an official newly arrived in the district. The friend was to call on the official in the evening, and to hear his report Wei arranged to meet him outside the Lung-hsing monastery at dawn on the following day.

Wei Ku was so anxious for his friend to succeed that he hardly slept a wink that night. Long before dawn he got up and set off for the monastery to await his friend's arrival. When he arrived the moon was still shining, and there on the steps of the monastery an old man was sitting, reading a book by its light. Wei Ku looked over his shoulder, curious to see what he was reading, but he could not make head or tail of what he saw.

'Please tell me what you are reading,' he said to the

old man. 'I have never seen writing like that before. When I was young I read widely, but there is no form of Chinese that looks like that. It isn't even Indian writing. What is it?'

The old man laughed. 'How could you be expected to recognize it?' he asked. 'This is no book from the world of men—it is the writing of the underworld.'

'If that is so, then you must belong to the underworld yourself,' said Wei Ku. 'What are you doing here?'

'I might ask the same question of you,' replied the old man. 'We of the underworld have to arrange the affairs of mortals, and how can we do that without visiting your world now and again? We are careful to come at times when there is no one about. Either I have stayed too late this time, or you are up and about too early: anyway, we have met, and there is an end of it.'

He was returning to his book, but Wei Ku wished to know more. 'You speak of arranging the affairs of men,' he went on. 'What particular affairs are you in charge of, may I ask?'

With a sigh the old man put down his book again. 'I look after people's marriages,' he said.

Wei Ku's face lit up with joy. 'Then you are just the man—I mean, spirit—I am looking for. My parents died when I was very young and before they had found a wife for me. I have been looking for one for years, but without success. Now a friend of mine is trying to arrange a match for me with the daughter of the new magistrate. Will I be successful this time?'

'Let me see,' said the old man, and thumbed through his book until he found the place he wanted. He studied the page for a moment, then shook his head. 'No,' he said. 'Your future wife is three years of age at this moment. You will marry her when she is seventeen.'

Fourteen years still to wait! Wei Ku was disconsolate,

but did not doubt the truth of what the old man said. Sadly he turned away. As he did so, he noticed a bag lying on the steps by the old man's side. It was open at the top, and seemed to contain reel upon reel of red thread. 'What do you use this thread for?' he asked.

The old man was once again immersed in his book and seemed to have forgotten Wei's existence. He now looked up again, and answered, with some impatience, 'This is the thread with which I tie together the feet of those who are destined to marry. Once I have tied them together with this thread they will become man and wife. It doesn't matter whether they are a thousand miles apart, whether one is rich and the other poor, whether their families are at daggers drawn with each other. At the appointed time, marry they must.'

'Then is there a thread tied to my feet?' asked Wei Ku. 'And can I see who it is at the other end?'

'I told you she is a child of three,' said the old man. 'But she is not far from here. It is dawn now—come, I will show her to you, and then perhaps you will stop pestering me.'

Wei Ku's friend had not turned up, and he realized the attempt must have failed. The old man picked up his bag, and Wei Ku followed him to the market-place, which was beginning to fill up with people. Presently, along came a peasant woman, ragged and dirty and blind in one eye. She pushed a barrow piled with vegetables which she was offering for sale, and on her back, wrapped in a bundle of tattered cloths, was a grubby child of two or three.

The old man pointed to the child. 'That is your bride,' he said.

Wei Ku was filled with disgust. 'If I kill her, that will end the matter,' he said.

'This girl is destined to enjoy wealth and high rank,'

said the old man. 'There is nothing you can do about it.' And with these words he disappeared.

Wei Ku went home. His mind was in a turmoil, and he was determined to escape from this degrading match. Finally he took a sharp knife and called for a servant. 'Tomorrow,' he said to the man, 'you will come with me to the market-place. There I will point out to you a baby girl. She is an evil spirit. If you kill her for me with this knife I will reward you with ten strings of cash.'

The servant believed his master's words, and the next day accompanied him to the market, the knife concealed in his sleeve. Wei Ku pointed out the girl to him, and then returned home, not wishing to witness the terrible deed that he had ordered. The servant went up to the one-eyed pedlar-woman, took out the knife and prepared to stab the child in the back. But just as he was about to do so, the little girl squirmed round and looked at him with wide-open eyes. The servant could not believe that she was anything but an ordinary child. Deliberately he aimed to miss her, and caught her only a glancing blow over the eyebrow with the blade of the knife. Then he managed to lose himself in the press of people and make his escape. He reported to Wei Ku that he had in truth killed the girl, and received his reward. Wei Ku was long plagued with remorse over his crime, but in the end succeeded in dismissing the whole affair from his mind.

He made many fresh attempts to find a wife for himself, but none of them came to anything. At last the matter was decided for him, years later, when Wei Ku was serving as assistant to the Governor of Hsiang-chou. This man was very highly impressed by Wei Ku's work, and on learning that he had no family decided to give his own daughter to him in marriage. Wei Ku was delighted that after all these years his loneliness was to be ended. Besides, the Governor's daughter was a most beautiful girl of seventeen, and as pleasant-natured as she was beautiful.

The young bride wore her hair in the elaborate style which was then in fashion, and kept it in place by means of hairpins skilfully worked and decorated with precious stones and kingfisher feathers. One of these pins was placed at her temple and the head of it covered her forehead just above the eyebrow. This ornament she never removed, not even when she washed or retired for the night. After some time this roused Wei Ku's curiosity and he determined to ask the reason for it.

'The ornament covers an ugly scar on my forehead,' his beautiful wife explained. 'When I was very small my governess took me to the market one day and a madman tried to stab me.'

'But—you were in rags!' exclaimed the astonished Wei Ku.

His wife was equally surprised. 'How did you know of this?' she asked.

'Wait,' said Wei Ku. 'First tell me—was your father very poor when you were a child?'

'The Governor of Hsiang-chou is not my real father,' replied his wife. 'My parents, like yours, died when I was a baby. I was cared for by my old nurse-maid until I was seven, when my uncle, the present Governor, adopted me as his daughter.'

'And your nurse-maid was blind in one eye!' said Wei Ku; and he told her the whole story of his meeting with the old man in the moonlight and his attempt to have her killed in the market-place. His wife marvelled at the story. She forgave him for the crime he had attempted, and during the whole of their long and happy life together neither of them ever referred to it again.

The Three Precious Packets

A POOR student whose name was Niu was once travelling to the capital for the examinations and stayed overnight in a village inn. It was the depth of winter, bitterly cold. Snow fell steadily from a leaden sky. Niu rejoiced in the warmth and comfort inside the inn. He sat toasting himself before a charcoal brazier and ordered hot soup and dumplings for his supper. While waiting for them he drove the chill from inside him with a bowl of hot rice-wine. Soon he felt a pleasant glow suffuse his whole body.

The brief winter daylight was almost gone when a man knocked on the door and was admitted stumbling into the inn. Clearly he was poorer even than Niu himself. His clothes were thin cotton rags, no proof against the keen weather. His cheeks were pinched and he moved stiffly as though half-frozen. His lips were too numb to allow him to speak, but he cringed before Niu, obviously pleading to be allowed to share the warmth of the brazier.

Niu quickly made a place for him, and the man

stretched his hands eagerly towards the glowing charcoal and began gradually to thaw out.

The landlord brought Niu's supper, the steam from which tickled the nostrils with a most delicious smell. Then he turned to the ragged newcomer and asked what he could do for him.

'A bowl of tea, and a place on the floor for the night,' said the man, and the landlord grunted and went off to fetch the tea.

Niu was filled with pity for this poor stranger. He pressed on him his own supper, and when the landlord came back ordered more for himself. The stranger wolfed down the soup and dumplings in no time at all, and Niu ordered noodles to follow.

'I have no money to pay for them,' said the man; but Niu insisted that he should accept them, and before long five bowls of noodles had disappeared in the way the soup had done. When they retired to rest Niu tried to give up his own bed for the stranger to sleep on, but the man would not accept it. He lay down on the floor at the foot of the bed and slept immediately. As if to prove his satisfaction with the supper he had eaten, he snored like a bull all through the night.

As dawn was breaking Niu was brought into wakefulness by someone gently shaking his shoulder. It was the stranger, who gazed at him with eyes full of affection. 'Please come outside with me,' he said. 'I have something to say to you.'

Wonderingly Niu pulled on his boots and followed the man out of the door of the inn. In the cold grey light the stranger turned to him and said, 'I am not a human being, but an envoy from the underworld.' Seeing Niu's look of fear, he added hastily, 'Don't worry, I have not come for you. But I am very grateful for your kindness to me last night. If you will lend me brush and ink and some paper I will give you something in return.'

Niu brought out these things from the inn, and the stranger told him to go away a little distance, while he went to sit under a tree. There he drew out from his sleeve a little book, and began to thumb through the pages. He stopped and copied out a few words on to a strip of paper, which he then folded into a packet. This he did three times, and numbered the packets, one, two, three. When he had finished he gave the packets to Niu with the following words, 'When you meet with the ordinary difficulties of life, do not think of these packets. But when you are faced with great trouble and the future looks impossible, then burn incense and open one of them. Only open one packet at any one time; and be very sure that you open them in the correct order, one, two, three. I must leave you now. Good-bye, and thank you again.' And he walked a few steps, then disappeared.

Niu put away the packets in his bag of books. He did not in fact place very much faith in what his strange friend had said, and soon forgot all about the incident. He went on to the capital and there awaited the examinations. But food and lodging in the capital were a great deal more expensive than in village inns. Before long Niu found he had very little money left and was forced to cut down his meals until he was almost starving. He earned a little by working as a scribe, but this interfered with his preparations for the examination, and soon he was at his wit's end.

It was then that he remembered the packets he had in his bag. He bought some incense and burned it, then took out packet number one as instructed, and opened it. Inside was a slip of paper bearing the command, 'Go and sit outside the gate of the Bodhi Temple'.

The Bodhi Temple was over ten miles away. It was a miserable day of cold and sleet, and Niu had hardly the strength to leave his pauper's lodging. Nevertheless he spent his last few coppers on the hire of a donkey and set out for the temple.

Barely had he seated himself on the frozen ground outside the gate when a monk came out to him. 'What are you doing, sitting there in this weather?' the monk scolded him. 'If you sit there much longer you will freeze to death, and then we shall get into trouble.'

'I am a scholar,' said Niu. 'All I want is to be allowed to spend the night on the steps of your temple. I will go away in the morning.'

'Oh, my apologies,' said the monk, 'I did not realize you were a scholar. We can furnish you with a night's lodging here if you will come in.'

In his quarters in the temple the monk asked Niu his name, then gave him a meal and had a long talk with him. At length he asked, 'Were you by any chance related to the late Magistrate Niu of Chin-yang?'

'He was my uncle,' replied Niu in surprise.

The monk asked him some questions about the magistrate's personal name and private affairs, all of which Niu answered to his satisfaction. Then he explained, 'Your uncle was a benefactor of this temple. But in addition to giving us a good deal of money he left three thousand strings of coins with me for safe keeping during the recent troubles. When he died I did not know what was to become of the money, but now you are here you might as well take it.'

From now on, of course, Niu was a wealthy man. He bought a fine house in the capital, kept servants and horses and carriages and lived altogether in fine style. Yet he was far from content, for his one great ambition

was to hold office as a magistrate, and this he could only do by passing the examination. He failed in the examination for which he had originally come to the capital, and three years later when the next set of examinations were held he failed again. He prepared for a third attempt, but with little confidence of succeeding. He was desperately unhappy, feeling that he would never attain his life's wish.

Then one day, as he was arranging his books in his study, there fell to the floor two soiled old paper packets, and he suddenly remembered the incident of the starving messenger from the underworld. He picked up the packet marked number two, wondering whether this could help him gain his ambition. Remembering the spirit's instructions he again burned incense, then opened the second packet.

He drew out a slip of paper, which again bore a command, 'Go and sit in Chang's restaurant in the West Market'.

Niu found Chang's restaurant easily enough. He took a seat upstairs and ordered tea. Behind his back was a curtain, which was used to screen off a private room. In this room two men were chatting, and Niu could not avoid overhearing their conversation.

'I have been amusing myself,' the first man was saying, 'by composing a little poem on the theme you set. It goes like this . . .' And he proceeded to recite with evident pride his own composition.

'Not at all bad,' said the other man when he had come to an end. 'But I am not sure I should pass you for that. You used a number of faulty rhymes. . . .'

And here he enumerated the mistakes that the first man had made in his poem.

The poet was rather annoyed by this criticism, and decided to get his own back. 'I am only afraid,' he said, 'that the essay subjects I have set will prove too easy. All, of course, are quotations, and even the dullest candidate will be able at once to recognize which works they come from. Take the first one, for instance. . . .' He recited the sentence which was to provide the subject for the first essay. Then he paused, clearly to allow the second man to identify the quotation. Equally clearly, the unfortunate man was unable to do so. He coughed and spluttered and tried to change the subject, but the first man was not to be put off. He went on to recite sentence after sentence, none of which his companion could recognize. At last both of them roared with laughter. 'Perhaps it's just as well,' they bellowed at each other, 'that we examiners don't have to sit the examinations we have set, or none of us would pass!'

Niu's delight knew no bounds. Before long he had overheard all the details of the questions which would face him when he attended the next session of the official examinations! He had an excellent memory, and when he reached home he sat down and copied out every word of what he had heard. He made his preparations for the examinations accordingly, and when the time came he had his answers complete in his head before ever he entered the hall. He passed with distinction, and to his great joy was given a high official post.

Niu earned a fine reputation as an official. His kindness to all in need or in distress became a legend. Wherever he went in the course of his career he carried with him the third of the precious packets; but it seemed as though his troubles were over and he began to think he would never need to open it.

But when he was still not very old he fell ill. One doctor after another bled him or prescribed medicines, prayers were said for him in the temples, but nothing seemed to have any effect. At last Niu decided that his only hope was to open the third packet. He burned incense, opened the packet and took out a slip of paper on which were written only three words, 'Make your will'.

Niu realized that his time was at hand. Without haste and without complaint he set his affairs in order and took leave of his family and friends. Then he died, peacefully, and mourned by all.

The Man Who Nearly Became Fishpaste

One day in the autumn of the year A.D. 759 the staff of a certain magistrate's court were taking their ease as they waited for lunch. Two men were sitting out in the courtyard playing chess. They looked up as a servant came in at the gate.

'Well done, Chang Pi,' said one of the two to the servant, and nodded his head in approval of the fish the man had bought for their lunch. It was a splendid fat carp, which Chang Pi carried by a string threaded through its gills. The chess-player turned to his companion and said, 'There'll be three or four pounds apiece out of that fellow!'

The other smiled and nodded. Idly he watched the servant cross the courtyard. Then he said in surprise, 'Do you know, I believe that fish knows what is going to happen to it! Look, there are tears running down from its eyes, and its mouth is moving—perhaps it's praying!'

And he called to those inside the residence to look at the praying fish. One officer was eating peaches from a bowl and two more were playing a game of dice as they sat in the hall, and they agreed it was a most extraordinary thing that a fish should look for all the world as though it were trying to talk to them. The servant Chang Pi smiled at their nonsense and bore his precious purchase proudly off into the kitchen to deliver to the cook.

Again the officers were distracted from their idle pursuits. This time it was by a long sigh which issued from a small side-room opening off the hall. In this room lay the assistant magistrate Hsüeh Wei. Several weeks previously this man had been stricken by a sudden fever. For seven days he had tossed and turned in delirium. At the end of this time he lay pale and exhausted on his bed, and those who attended him thought that his end had come. There was no real sign of life; and yet there was a little warmth about his heart, and so there could be no question of burying him. For twenty days more he had lain there, watched night and day. And now this sigh had come from him. His colleagues rushed into the room. Tears of joy streamed down their cheeks as they saw him sitting up in bed, to all appearances in the best of health.

'How long have I been away from you?' asked the man who had been so sick.

'More than twenty days,' came the reply.

His next question startled them: 'Have you been eating fishpaste?'

'We were just about to eat some,' they answered. 'But how did you know that?'

'Wait a minute. Did you send the servant Chang Pi to buy the fish?'

More mystified than ever they replied that they had. Hsüeh Wei turned to the servant: 'You went to the fisherman Chao Kan, who had caught a large carp. But he hid it in the reeds and offered you a small one.

Then you searched and found the big carp, which you brought back with you. When you returned to this residence, these two gentlemen here were playing chess, those two gentlemen were playing dice, and that gentleman was eating peaches. You told them how Chao Kan had tried to hide the big carp, and they gave orders that he should be beaten. Then you delivered the carp to the cook, who was delighted with it and killed it. All this is true, is it not?'

All agreed that Hsüeh Wei had described exactly what had happened.

'But how did you know all this?' they asked him.

'That carp,' said Hsüeh Wei, 'was me.'

And he proceeded to tell them his story.

'When first I fell ill,' he said, 'I had a raging fever. I felt I would do anything to obtain some relief from the heat. In the end I took my stick and went out for a walk.'

Observing the surprise on the faces of his audience—for he had not left his bed since the illness began—he went on, 'It did not seem like a dream, and yet I suppose it must have been. I left the city and made for the hills, feeling as happy as a bird released from its cage. When I grew tired of walking in the hills I came down again and followed the bank of a river. The water was clear and sparkling in the sunlight. Eventually I came to a pool formed by the widening of the river. Here the water was still and deep, cool and inviting. I could not resist the temptation—in the end I took off my clothes and plunged in. I have not swum since I was a boy, although I was an expert then; but as I glided through the water I realized that this was just what I had been longing for. I began to regret that I was not a fish, so that I could swim with perfect ease. "If only I could become a fish!" I said to myself.

'Immediately a fish which was swimming by said to me, "If that is really what you would like to be, it's quite easy to arrange."

'The fish swam off; and before long a strange creature appeared. It had the head of a fish but the body of a man. It was several feet tall, and rode on the back of a giant salamander. Several dozen fish swam in its train. This creature told me of a proclamation by the Lord of the Rivers. He said that creatures of land and water must normally follow different ways. But since I had tired of land and wished to find freedom in the waters, my desire should be granted and I might take the form of a golden carp. But I must be careful neither to cause damage by stirring up the waves and overturning boats, nor to risk my safety by swallowing bait. By doing these things I should bring disgrace on my kind.

'I now discovered that I had the form of a handsome golden carp. I shot through the water with ease and grace. I played in and out of the waves, or dived down to the still depths, roaming just as I liked through the rivers and lakes. Only, since I had been given that one particular pool as my home, I must return there every evening.

'Suddenly one day I began to feel hungry. I could find nothing to eat. I followed a boat downstream, until at length I saw the fisherman Chao Kan dangling his bait in the water. I caught the scent of it—it was delicious. Although I clearly remembered that I must not take bait, somehow or other I found myself nosing it. Then I said to myself, "I am a man, not a fish. I have only taken on the form of a carp for the time being. I certainly must not allow myself to swallow Chao Kan's bait when I know perfectly well that is a hook hidden in the middle of it." And so I swam away

a little. But before long I was hungrier than ever. I began to reason with myself: "Look here, you are an official, you're only playing at being a fish. Supposing you do swallow the hook—Chao Kan isn't going to kill you!"

'Confident that Chao Kan would arrange for me to be taken back to this residence, I swam up to the bait again and swallowed it, hook and all. Chao Kan, who is a very clever fisherman indeed, hauled me in to the bank. I called out to him, but he didn't seem to hear me. Instead, he threaded a cord through my gills—that is to say, my cheeks—and tied the other end to a clump of reeds.

'Then along came the servant Chang Pi, who said, "The officials would like a fish, and they want a big one."

'"I haven't caught any big ones," Chao Kan lied. "You can have a small one of ten pounds or so."

'"It has to be a big one—what use is a little one?" said Chang Pi; and he looked in the reeds and found me and lifted me out.

'"I am the assistant magistrate," said I to Chang Pi. "I have merely changed myself into a carp for the time being so as to swim about more easily. Why don't you treat me with the respect due to your superior?"

'But Chang Pi took not the slightest notice. He began to carry me back here. I shouted and stormed at him but he didn't even look at me. As we entered the courtyard I saw two of you sitting there playing chess. I called out at the top of my voice, but there was no response. Only, one of you turned to the other and said, "There'll be three or four pounds apiece out of that fellow!" Then I saw two more of you playing dice, and another one eating peaches, and all of you were delighted with the size of me. Chang Pi told you how Chao Kan had tried to hide me and substitute a smaller fish, and you ordered him to be beaten.

'In desperation I shouted to you, "Gentlemen, I am your colleague. How can you have the heart to kill me?" I wept and sobbed, but it was all to no avail—you handed me over to the cook. He laid me on his slab and took up his knife, and even as I was pleading with him to spare my life, down came the knife and off rolled my head. I made one dash back to my own body—and here I am.'

All remembered how they had noticed the fish's mouth moving—Chao Kan when he had hauled it in,

Chang Pi when he had lifted it up, the officials playing chess and dice, the cook in the kitchen; yet none of them had heard a sound.

But not one of them was ever able to touch fishpaste again for the rest of his life.

A Shiver of Ghosts

Ghosts have in many ways an easier time than living people. They need to spend far less effort on providing food and drink, travel presents very few problems to them, and as far as accommodation is concerned, they are quickly suited. Any nice dark, cheerless house or hut or grave-mound will do, provided it is inconveniently situated, reasonably airless, and good and dank.

When they are not busy haunting the world of the living, ghosts dwell in a land which is all their own. Centuries ago, one man visited the Country of the Ghosts and came back to tell the tale. He was a seafaring man, a trader, and like all of his kind he was stout-hearted and full of a lively curiosity about the many lands he visited across the sea.

His junk, on this strangest of all his voyages, was three weeks out of his home port in Shantung. For days on end the huge mat sails had creaked and strained before a steady breeze. But now a sudden squall came up. In a world of heaving seas and flying spray which seemed to cut out the sky above, the junk was twirled and hurled this way and that until neither the trader nor his helmsman had any idea how far she was off course. When the storm at last abated they could only make the haziest guess at a bearing and hope for a speedy landfall.

After many days the welcome moment came. A shout of 'Land ho!' rose from the look-out, and fixing his gaze the trader made out the blue line of land low on the horizon. As the ship sped nearer, details could be distinguished. The vista was full of promise. There were gentle green hills and darker valleys which must be wooded. Now it became clear that they were heading towards the mouth of a wide river: and at last the trader's heart gave a joyful bound when he traced the brown line of a city wall against the foot of the hills.

Every man of the crew was asked if he could name this place. None had seen it before. But a white-haired old sailor said, 'In fifty years of seagoing I have never been here before. But if we are where I think, then this must be the Country of the Ghosts we've all heard tell of.'

The crew paled at these words, and man after man pleaded with the trader to turn back. But the trader laughed, 'This is old women's talk! Those hills and that city wall are solid enough for me—or are those ghost-sheep, too, up on that hill-side!' And he gave the order to land.

While a party was detailed to water the ship from the near-by river, the trader himself with an escort of four men set off in the direction of the city. From the shore they found paths which led between plots of beans and paddy, trim and green as any to be seen in China. Soon they came upon an old man hoeing near their path. The trader called a polite greeting, but the old man took no notice. Thinking he must be deaf, they continued as far as a cluster of huts around which hens and chickens pecked at the ground. A young man came out of one of the huts and walked down the path towards them, a spade over his shoulder. They made way for him, and he walked straight past without giving them a glance. 'He must be blind,' said one of the men, 'but how well he walks for a blind man.'

But as they went on and met and greeted men, women and children without once finding any response, they realized that these people were neither deaf nor blind—to each other; they simply did not perceive the presence of the strangers. What the old sailor had guessed was the truth—they were in the Country of the Ghosts.

Now the city wall rose before them. Their path led straight to the gate, which like any gate in China was kept by an old gate-keeper. Aware by now that they could go where they pleased without hindrance, the five strangers walked past him into the city. It was a pleasant place of broad streets and well-kept houses, quiet alleys and crowded market-places. The trader and his men might have thought themselves in any town in China, but for the fact that no one took the slightest notice of them, and once or twice, expecting to be jostled on the busy street, they found themselves passing straight through human forms which might have been shaped out of mist.

Emerging into an open space in the centre of the city, they saw across from them a magnificent edifice. Along the whole façade carved crimson pillars supported a sweeping roof of brilliant yellow tiles. It could only be the palace of the ruler of the land. Nothing daunted, the trader led his men past the guards at the splendid gateway, through covered walks where singing-birds were caged and across the marble floors of courtyards where fountains played and flowers charmed the eye.

From a large lofty hall came the sound of music and laughter. Within a feast was in progress. The ministers of the court were seated at long tables loaded with sumptuous dishes and elegant wine-jars, whilst in the open centre of the hall a bevy of dancing-girls swayed their brocaded sleeves. In the centre seat of the top table sat a being robed in splendour and with a countenance filled with majesty even in this moment of ease and laughter. The trader made his way not between but through the dancers, as though he were passing through the patches of light and shade of some sunlit forest, until he stood a few paces before the King. He prostrated himself on the floor as he would have done before the Dragon Throne of China, then rose and gave most reverent greeting to His Majesty. But His Majesty gave no more sign of being aware of the trader's presence than any of his subjects had done. It seemed almost as if he and all his people were simply some elaborate puppet-show: and yet they were eating and laughing, singing and dancing, full of activity. Now the trader could not restrain his curiosity, and he felt impelled to make the most of this strange opportunity to observe royalty at the closest range. He made his way right up to the table, leaned across and peered straight into His Majesty's face. But he could only gaze for a few brief seconds: for the King suddenly slumped sideways in his seat and the wine-bowl he was holding smashed to the floor.

Instantly all was in uproar. Attendants raised the King carefully from his seat and carried him into a side-room while the court physician was sent for. He hastened to the scene, examined the King and then said, to the great relief of all (not least the trader), 'It is not serious. His Majesty has merely fainted. But I cannot discover the reason. There is something mysterious here. Summon the court magician.'

In a trice a venerable old man entered the sick-room, his long black robe embroidered with mystic symbols. He in his turn examined the King. Then he addressed the ministers who had watched him in anxious silence, 'His Majesty will soon recover, but the cause of his distress must be removed at once. His Majesty has been overcome by the breath of the living.'

'The breath of the living!' repeated the ministers in astonishment. 'But where, how, in this country of ours?'

'A living man is in our midst,' said the magician. 'He has reached our country by chance, and means no mischief. But he approached His Majesty too closely, and thus caused this faint. Let us treat him with kindness but ask him to leave our land.'

None questioned the word of the court magician. At once a feast was set out in a separate hall, where the most skilled of the musicians and dancing-girls were bidden to perform. The Chamberlain addressed the empty air in ringing tones. 'Man of the living, honoured guest!' he announced. 'We of the Country of the Ghosts are grateful for the favour of your visit. We invite you now to partake of this feast we have prepared for you. When you have eaten and drunk to your contentment, you will find horses and carriages waiting outside the hall to take you in safety back to your ship.'

The trader and his companions had no wish to cause further anxiety. They ate and drank their full, and all the while the court magician uttered invocations and prayers for their safe departure, while the ministers watched and marvelled that the food and wine grew steadily less, though they could not perceive the feasters. When they had done, the travellers found outside the hall the carriages which had been promised. Smoothly and safely they were carried back to the shore. On the seat of his carriage the trader found a lacquer box in which was bar upon bar of heavy, gleaming gold, and this box he carried back on board the junk. But as he left the shore he felt the box grow suddenly lighter. Opening it again he found that the bars he had thought were gold were nothing but yellow paper—it was the paper money men burned at funerals to provide for the dead. Raising his eyes for one last look, he could see no horses nor escort nor any sign of people, only the hills and valleys and the river flowing down to the sea.

A stiff off-shore breeze was blowing. The junk made sail and with no further mishap returned to port in Shantung, where the trader told his story far and wide. But from that day to this, no one has ever again seen on the horizon the hills of the Country of the Ghosts.

That is not to say, however, that no ghost since then has been seen in the land of the living. They have been encountered in all kinds of ways, some of them the most matter-of-fact imaginable. There was Sung Ting-po, for instance, who met a ghost in the early hours of the morning when he was walking to market. When he asked it who it was, it replied, 'I'm a ghost.'

Sung Ting-po was a stalwart young man and not given to panicking, and so when the ghost in turn asked who he was he replied, 'I'm a ghost too.'

'Where are you going?' asked the ghost.

'To market,' answered Sung.

'So am I,' said the ghost, and they walked on together for a mile or so. Then the ghost began to complain that their progress was too slow, and suggested that they took turns at carrying each other in piggy-back style. It was agreed that Sung Ting-po should have the first ride. But when it had carried him a few paces the ghost said, 'How heavy you are! You're much too heavy to be a ghost.'

'Well, you see,' Sung explained, 'I haven't been a ghost very long, and my body hasn't had time to lighten yet.'

This seemed to satisfy the ghost. Soon came its turn to ride, and Sung found there was no weight to the ghost at all.

So they continued, carrying and being carried in turn. Eventually Sung Ting-po remarked, 'Being a very new ghost, I don't know much about it yet. What is it that we ghosts are most afraid of?'

'There's only one thing a ghost has to fear,' came the reply, 'and that is to have a living man spit at one.'

They came to a stream, which they decided to wade across separately. The ghost crossed first, making no sound in the water. But when Sung Ting-po tried, he made a tremendous splashing which revived the ghost's suspicions. 'Why do you make such a noise?' he asked sharply.

'It's as I say, I really have very little idea of how to move about as yet,' said Sung. 'We new ghosts have so much to learn!'

Dawn was breaking as they entered the market-place. The ghost was on Sung Ting-po's back at this point, but as it tried to get down Sung gripped it tight. It squealed and yelled but he wouldn't let go. Grasping it firmly by the arms he brought it to the ground. But the ghost had seen the various animals waiting in the market-place to be sold. As it touched the ground it changed into a sheep, no doubt with the idea of running away and losing itself in the pens near by. Remembering what he had learnt, Sung Ting-po spat at the sheep so that it would not be able to resume its ghost-shape. Then he led it firmly by the neck up to a butcher, who gave him fifteen hundred cash for the sheep without a second's hesitation. This is probably the only case on record of the sale of a ghost for fifteen hundred cash.

Another ghost-catcher, Lo Ta-lin, went about his task in rather a different way. Lo was a pedlar, not at all quick-witted like Sung Ting-po, but tall and swarthy and immensely strong. He had shoulders as strong as a

horse; in fact, once when a friend's horse went lame he lifted it up in his own two arms and carried it back to its stable.

He was a rough fellow, and none too modest. He used to boast that he feared neither man nor devil. Only one thing he regretted, and that was that he was a poor man. He would dearly have loved to take a wife, but he had no money to buy presents and pay for the wedding. One day he was complaining to his friends in his usual surly way that a pedlar's job was no better than a punishment and that he would go down to the grave without a string of cash to his name. They began to grow impatient with him, and at last one of them burst out, 'You're always grumbling how little you earn, and always boasting how brave and fearless you are. You say you fear neither man nor devil. Well, then, would you be afraid of a ghost?'

Lo Ta-lin roared with laughter. 'I'd like to see a ghost with a back as broad as mine,' he sneered.

'Then there's one way you could earn yourself a fortune,' said his friend.

'And who's going to give me this fortune?' asked Lo.

'Old Moneybags Wang, if you'll rid the Willow Lodge of whatever it is that haunts it,' his friend replied.

The others gasped at the boldness of this suggestion. It was true, old Wang would give a lot of money to be able to live in the Willow Lodge again. It was a fine mansion set in a grove of gracious willows. Wang had bought it cheap after it had stood derelict for years. He had renovated it and moved in with his family. But the next morning two of his servants were dead, mysteriously, unaccountably dead. Wang moved straight back to his old house and advertised that the Willow Lodge was to let. Three tenants in turn had taken the mansion, and each was found dead the morning after he had moved in.

Lo Ta-lin knew all this. But his eyes glinted at the thought of the reward old Moneybags might give to whoever would rid him of this ghost. He went straight round to Wang's house to put the proposition to him. Wang's servants tried to turn away this coarse fellow who came bursting in on their master, but when Lo Ta-lin flexed his huge muscles they ceased to try very hard.

Wang received Lo's offer with delight. 'There'll be a hundred strings of cash for you if you rid the Willow Lodge of this ghost,' he promised. Lo's eyes gleamed brighter than ever. 'And I've a nice little house by the market-place,' Wang added. 'I'll give you that for your bride.'

Lo was overjoyed. Then slowly a suspicion began to form in his mind. 'How do I know that you'll keep your word?' he asked.

Wang laughed. 'Very well then, we'll make a contract.' And he ordered his servants to bring paper, brush and ink, and made out an agreement as follows: 'In return for the service of clearing the Willow Lodge of whatever ghost, spirit or demon is at present haunting the said mansion, I, Wang Hsin-hung, do hereby undertake to give to Lo Ta-lin one hundred strings of cash together with one house in Mule Alley off the market-place.' Wang signed the document and passed it to Lo Ta-lin. Lo could no more read it than fly, but he stared at it, waggling his head from side to side in pretence, then stubbed his great thumb on the ink-slab and made his mark right in the centre of the paper.

That very evening Lo made his preparations. These were very simple. They consisted of buying a number of large candles, a pound of garlic and a few pints of wine. He ground the garlic, mixed it into the wine and drank till he was just a little fuddled. Then he lit one of the candles, drew back the bolts on the outer gate of the Willow Lodge, and entered.

A large crowd had accompanied him as far as the grove of willows. A sigh went up as they saw him enter the gate. Then one man said, 'I know Lo Ta-lin's plan all right. He'll wait till we've all gone home, then he'll come sneaking out of that gate again. He'll hide in the grove here all night, and then sneak back in when morning comes. Then when we come and find him there he'll cook up some tale of chasing off a whole army of ghosts at midnight.'

'In that case,' said another, 'perhaps we should make

sure he doesn't get out till morning.' And he bolted the gate again from the outside.

Lo Ta-lin heard the bolts crash to behind him, and grunted. The light of his candle (for there was no light from the sky) made a circle round him in the courtyard. Matted weeds trailed across the flower-beds, grass thrust up between the paving-stones. The inner gate was overgrown with creepers and brambles which Lo had to tear away. Inside the mansion dust lay thick in every room. It flew up as he moved, choking him, and spiders' webs clung to his face at every step. It was a place of horror. But at last Lo Ta-lin, nothing daunted, came to a room where there was no dust, no spiders' webs. And yet the room did not look lived in: only there was a bed against one wall, complete with bed-curtains and sheets no doubt left by the last tenant. Lo Ta-lin sat on the bed, snuffed out his candle and waited. He did not lie down, he did not close his eyes. He sat very still, and waited.

In the darkest and stillest hour of the night the ghost came. There was a mighty crashing outside the room, the door was flung open and the ghost rushed in. It ran round and round inside the room. Lo could only barely make it out in the gloom, for it seemed to be dressed in dark clothes and its face and hands were no lighter than its garments. Straining his eyes Lo waited until the ghost was near his bed. Then he sprang out, full in its path, and seized both its arms in his vice-like grip. The ghost, stopped and held in its wild career, struggled to free itself but to no avail. Its arms were pinioned to its sides and it could do nothing.

Man and ghost stood face to face. Then the ghost began to blow in Lo Ta-lin's face. Its breath was ice-cold, fearfully cold. Lo turned his head away. But the ghost's breath continued to blow on his neck, which soon began to ache intolerably. It was just as though a knife were being driven into it. Finally Lo could bear it no longer. With a great effort he twisted his head back to face the ghost, and began to blow in his turn.

Now Lo's own breath was flavoured with a good pound of garlic, and a ghost is no more capable than anyone else of standing unmoved while a pound of garlic is breathed into its face. The ghost turned its head away, and every time it tried to straighten up it met the nauseating stink of garlic. Lo kept it up until he ran out of breath. The ghost now seized its chance and blew again into Lo's face. So the strange, still battle raged between them, first one and then the other blowing on his adversary.

The long night drew painfully to its end. From somewhere in the world outside the first cock crew. This was the moment Lo Ta-lin had been waiting for. As the cock crew he grasped the ghost's arms even tighter than before, anxious lest it should disappear. But the ghost did not vanish. Instead, it began to shrink. There in Lo's arms it shrank and shrank. No longer was it able to blow in Lo's face. To his surprise, Lo felt the ghost's body growing harder and harder in his grasp.

At the crack of dawn Lo Ta-lin's friends had gathered in the willow-grove. Now, in the morning light, they drew back the bolts and entered the mansion. Through the rank gardens and the dust-choked rooms they advanced, filled with the dread of what they might find. Their relief knew no bounds when they saw Lo at last, standing unharmed in the middle of the room. In his arms he was clutching a plank of wood. As they entered he let it go, and it clattered to the floor. They examined it curiously: it was a plank of wood from an old coffin. 'There is your ghost,' said Lo, 'burn it and have done.'

They took the plank out and burned it. It gave off so

foul a smell as it burned that no one could go near it. Old Moneybags Wang lost no time in honouring his promise to Lo of a hundred strings of cash and a house, and Lo Ta-lin lost no time in finding himself a wife and settling down with her in Mule Alley. But every time the wind blew, for the rest of his life, he suffered pains from his wry neck, which he was never able to cure.

The important thing is not to be afraid of the ghosts one meets. It is only men's fear that gives them their power. An instance is the case of a man named Ts'ao who once visited a friend in Yangchow. It was the height of summer, and Ts'ao was greatly taken with his friend's cool and airy library. 'What a pleasant place this is,' he remarked. 'I wonder if I might be permitted to have my bed made here by the window?'

'You are most welcome to use my modest library during the day,' said his host, 'but I am afraid I cannot allow this room to be used for sleeping.' He was reluctant to give any reason for this, but Ts'ao at last obliged him to explain: 'My library, I am sorry to say, is haunted.'

'Haunted!' exclaimed Ts'ao. 'By what?'

'By the ghost of a maidservant who hanged herself from that beam many years ago.' And his host pointed upwards.

But Ts'ao's desire to sleep in the library was only enhanced by this fascinating news. His host yielded, and a bed was made by the window. That night Ts'ao read until late. Hardly had he put down his book and turned in than the ghost duly appeared. Ts'ao was lying with his eyes directed towards the door. Through the slit between the door and the jamb a shape appeared, no thicker than a sheet of paper. Before Ts'ao's delighted eyes it spread out into the form of a girl. She moved to the centre of the room and stuck out her tongue at him. But it was no ordinary gesture of rudeness. Her head jerked, her mouth opened and the tongue came rolling down in the manner of one who has been hanged.

'That is an interesting trick,' said Ts'ao. 'Please do it again.'

Seeing that she had failed to scare him, the ghost paused for a moment. Then she took off her head and put it on the library table.

'Since I am not afraid of you with your head on,' said Ts'ao calmly, 'why should I fear you with it off?'

The ghost had obviously exhausted her repertoire, and disappeared. The next morning Ts'ao told his host all that had happened. His host smiled a little uneasily and begged him to remove his bed from the library. But Ts'ao pleaded all the more urgently to remain there.

That night, at midnight, the ghost appeared again as before. But as soon as she had taken shape in the middle of the room she saw that it was the same man lying there. The ghost spat on the floor and said in disgust, 'Oh, it's that stubborn fellow again'—and promptly disappeared.

And yet, wrongdoers of all kinds have good reason to be afraid of ghosts, who have their own means of securing justice.

A certain lady, a member of a wealthy family of Shang-ch'iu, had been left a fine property by her late husband. She was a virtuous woman and was respected by her friends for her vow never to marry for a second time. But she had the misfortune to attract the envy of her dead husband's brothers and especially of their wives, none of whom was able to rival the quiet elegance of the widow's dress and the number of her servants. So these malcontents plotted how they could lay their hands on the widow's possessions. They started evil rumours against her. It is well known that a malicious tongue is as sharp as any sword, and at last the day came when an accusation was brought before the local magistrate that the widow had committed adultery with a man who had since left the district.

Still all would have been well (for they could find no proof of something that had never happened) had not the magistrate been a man named Ku. For Ku was a corrupt man and accepted heavy bribes from the widow's brothers-in-law. When the widow entered the court she knew that her cause was hopeless, for Ku was pledged to convict her in return for the favours he had received from her accusers. Seeing no way out, the widow, as she stood there in court, drew a knife from her sleeve, stabbed herself and fell lifeless at the magistrate's feet.

The case at once turned into a scandal. Higher authorities began an inquiry and discovered that Ku and all his underlings had accepted large bribes. Ku was dismissed from office and stripped of his illegal gains. He returned to his family home in Soochow, where he had sufficient property to keep him in reasonable comfort.

Many years after this distressing case was over, a merchant by the name of Yang chanced to pass through Shang-ch'iu. It was late in the day, and he sought shelter in a handsome inn. Now this inn was established in no less a place than the widow's old mansion, which had been sold after her suicide in court. The merchant Yang, of course, knew nothing of all this. He was disappointed to find no room for him that night. 'Have you nowhere at all that I could spread my bedding?' he asked the inn-keeper.

'Every room I have is full,' replied the latter. 'Except one, that is, and you can't sleep there.'

'What room is that, and why shouldn't I sleep there?' Yang asked.

'It is a room I should have pulled down years ago,' said the inn-keeper. 'Every guest who has slept in it has complained of its being haunted.'

'Well, if that's all that's wrong with it, I'll take it,' said Yang with a laugh. 'I promise you I won't complain.' And against his will the inn-keeper was forced to give him lodging.

The room was comfortable enough and Yang slept until the small hours, when he awoke to find a woman standing before his bed. She was pale and there was blood on her dress and on her hands—it was the ghost.

The woman leaned over him, but Yang told himself not to be afraid. She addressed him in a thin voice; 'Are you a man from Soochow?' she asked.

Yang replied that he was, and she went on, 'I suffered a wrong, and you can avenge it if you will. No harm will come to you.'

Yang said in surprise, 'I am a merchant, not a magistrate. How can I avenge your wrong?'

'Take me with you when you return,' said the ghost. 'That is all I ask. If you will do this I shall reward you for your pains.'

'I am at your service,' said Yang. 'But tell me how I am to take you with me.'

The ghost gave him her instructions: 'When you leave on your return journey, call me. Call me again, just quietly, whenever you board or leave a boat or pass under or over a bridge. When you reach Soochow, I want you to call at the house of the former magistrate

Ku. Take an umbrella with you—I shall hide in the umbrella. That is all you have to do.' Then she went on, 'The reason I have never left this room is that there is a box of jewels plastered into the wall in the south-east corner. I want you to have it in return for your kindness to me.'

With that the ghost vanished. Yang turned over and went to sleep again. In the morning he chipped away a little of the plaster from the wall where the ghost had indicated. There, sure enough, was a small box filled with the most priceless pearls.

Yang followed the ghost's instructions to the letter. He called her in a low voice when he began his homeward journey, and again every time he boarded or left a boat or passed under or over a bridge. When at last he reached Soochow he bought an umbrella and made straight for the house of the former magistrate, Ku. It so happened that Ku had invited some strolling players along that day to entertain a few guests, and Yang was able to slip unnoticed into the gardens.

For a little while Yang watched the scene of merriment as the guests roared at the antics of two comic warriors. Then suddenly a scream rang out, chilling the blood of all that merry crowd. It was Ku, who a minute previously had been drinking and laughing with his friends. Now he leapt to his feet and stood there trembling, his eyes glaring out from an ashen face, the wine-bowl fallen from his nerveless hand. At last he raised his hand and pointed before him. He found his voice, a high screech, 'Take her out! Take her out! That woman! There is blood on her dress and a knife in her hand. I know who she is! Take her out!'

Servants came running, and all present stared in the direction of Ku's outstretched finger. But no one, not even the merchant Yang who still looked on in horror, could see anything there at all. Ku collapsed quivering into his chair. He was revived with strong wine, but the party was at an end and soon the guests were taking their leave.

That night, the magistrate Ku hanged himself from a beam in his own bedroom. And the room in the inn at Shang-ch'iu was haunted no more.

Magicians of the Way

BUDDHIST priests shave their heads. But often in a Chinese painting you will see a man, perhaps seated in meditation, whose pious attitude contrasts with his ragged clothes and the mass of unkempt hair on his head. He also will be a holy man, but of the Taoist belief, one who follows the Way or *Tao* of the ancient sage Lao Tzu. Many such men believed that by purifying themselves, or by eating magic herbs or concocting some secret pill or elixir, they could increase their power far beyond the limits of the ordinary human body. Perhaps some of them did so. At any rate, ordinary people heard enough stories about them to believe them the most skilful magicians in the land.

There was the matter of sleep, for instance. A Taoist named Ch'en T'uan, realizing that sleep strengthened both body and mind, set himself to perfect the art. He holds the world record with a sleep of eight hundred years. But he was not always left undisturbed for so long. He lived in seclusion on a mountain called T'ai-hua-shan. One day he was seen to go down the mountain, and although several months passed he did not return. Other holy men who lived on the mountain told themselves that he had gone to live elsewhere, and returned to their meditations.

Winter drew on, and the stocks of firewood that had been made ready against the cold began to dwindle. The day came when one man was nearing the end of the pile in his woodshed. He took hold of a long thin log and pulled—and discovered that he was grasping a human leg! Horrified, he removed the remaining logs from the top. Then he roared with laughter—to see Ch'en T'uan sit up, rub his eyes, stretch himself and begin to brush the shavings from his person.

On another day, much later, a farmer was scything grass for hay on the lower slope of the mountain. He came to a break in the ground where the dried-up gully of a stream ran down, and there in the gully he was saddened to see a corpse lying. 'Poor fellow,' thought the farmer, and he stooped to take a closer look. Grass and weeds were growing in the soil which had drifted on to the body, and a lark had built its nest between the feet. The farmer was moved to pity. He decided that he would bring a cart to take the corpse away and give it a decent burial. But at this point Ch'en T'uan woke up. Opening wide his eyes he said, 'I was just enjoying a most pleasant nap. Who is this that has disturbed me?'

The farmer recognized the great sage of the mountain. He apologized most humbly for having spoilt the master's rest, and then returned to his scything while Ch'en T'uan strolled slowly off to breakfast.

Liu Ken was a man who devoted himself to Taoism on the mountain Sung-shan. There he acquired many marvellous powers. One of his discoveries was the secret of youth. It was said that when he was over a hundred years old he looked like a boy of fifteen.

Nor did he benefit himself alone. Once a plague broke out in the city of Ying-chuan. Not even the family of the Governor himself was spared. But fortunately the Governor had a high regard for the Way, and when he heard that Liu Ken was in the vicinity he sought his help. Liu Ken at once gave him a piece of paper bearing strangely written characters. It was a charm, which the Governor must paste on the door of his residence at night. The Governor did so: and when morning came, not one member of his family showed any further trace of the dread disease.

But before long Ying-chuan had a change of Governor. The new man, an official named Chang, was very different from his predecessor. He sneered when his subordinates told him of the deeds of Liu Ken. When they persisted with their stories Chang grew angry. He determined to show up this man of marvels, and sent out runners to arrest him. News of this spread quickly through the city. When the runners left the Governor's residence they found the streets blocked by indignant crowds. They themselves had little heart for their task and turned back. But the Governor, nothing daunted, sent troops to break through the crowd and bring Liu Ken to court for questioning.

At length Liu Ken, barefoot and in rags, stood below the gilded chair of the Governor. The court had been cleared of all but the troops, who looked on impassively, determined not to risk their necks by crossing their stern superior. The Governor leant down and addressed Liu Ken in a voice of thunder.

'Are you a magician?' he roared.

Liu Ken's reply showed neither pride nor deference. Calmly he answered, 'Yes.'

'Can you bring down the spirits from the Next World?' the Governor continued.

'I can.'

The Governor was delighted. This was really too easy. This charlatan had fallen straight into his trap. Louder than ever he roared, 'Then let me see some spirits, now in this very court, or you will be punished as an impostor and a swindler.'

Liu Ken glanced to one side, where lay the instruments already prepared for his punishment: the boards for squeezing his ankles, the heavy bamboo rod with which he would be beaten. Then he turned back to the Governor, who was amazed to observe a quiet smile on the holy man's face.

'Nothing could be easier,' said Liu Ken softly. He asked for brush, ink-slab and paper. When a guard had passed them to him he made a few deft strokes on the paper, which he took over to a brazier and committed to the flames. No sooner had the last corner of the paper blackened than the court was filled with the din of trampling hooves and the clatter of armour. The wall at the far end of the court had swung open, to admit a great troop of riders armed with swords and spears. They crowded, four or five hundred of them, into the hall. Then their ranks parted, to reveal a cart in which knelt the two prisoners they were escorting.

Liu ordered the prisoners to be brought out of the cart and led up to the Governor. One was an old man, the other clearly his wife. The old man raised his head and addressed the Governor in tones of stern reproof:

'To think that you, our son, should have brought such disgrace upon us! Do you think your duty to us ended with our death? Or that you can give offence to a holy man without involving us also in your punishment? How can you dare to look us, your dead parents, in the face?'

Ashen-faced, the Governor hastened down from his seat and fell to his knees, weeping, before Liu Ken. He begged him to have mercy, to release his parents.

Liu Ken gave an order, and the Governor looked up to find the court empty save for the holy man and the guards. The far wall was as solid and substantial as ever it had been. And never again did the Governor Chang venture to doubt the powers of those who follow the Way.

Another magician who came up against the authorities was a Taoist named Tso Tz'u, who played a whole series of tricks on the august Duke of Wei. It all began when Tso Tz'u was walking one day along a highway and saw coming towards him a string of a dozen or so coolies, each one laden with a huge basket. Straw packing covered the top of each basket. Tso Tz'u stopped the little procession.

'What have you got in those baskets?' he asked.

'Oranges for the Duke of Wei,' answered the leading porter.

'Oranges!' exclaimed Tso Tz'u. 'How nice for the Duke! But what heavy loads for you poor fellows! Let me help you a little of the way.' And so saying he took the basket from the leading coolie and hoisted it on to his own back. He carried it for a mile or so, then returned it to the first coolie and took over the load of the second. In this way he went right down the line. Now every porter when he received his load back found it lighter than before, and each wondered whether this was merely because of the rest he had had, or whether the Taoist had worked some magic charm for his benefit. When Tso Tz'u had reached the end of the line he took his leave of them, and in due course they reached the Duke's palace and delivered the oranges into the storehouse.

A day or two later the Duke held a small banquet for some visiting noblemen. When all the rich dishes had been disposed of, servants brought in silver dishes laden with luscious golden fruit, and the Duke turned to his guests: 'A rare pleasure, gentlemen! Sweet oranges, freshly brought from the south!'

The mouths of the guests watered as they contemplated the glowing golden fruit. But the first guest to cut one open with the silver knife he was given gaped in dismay—for inside the inviting skin was nothing but empty air. He cast a furtive glance towards the smiling Duke, and wondered fearfully what kind of malice such an insult as this might portend. Then to his surprise he noticed his neighbour looking up in exactly the same way, and then——

'Ancestors bear witness!' roared the Duke in a voice which set all the silver rattling on the tables. 'STEWARD!'

The Duke cursed the steward and his ancestors, and the steward cursed the storekeeper and *his* ancestors, and the storekeeper sought out the chief porter and cursed him and *his* ancestors: for every single one of the whole consignment of oranges was nothing but an empty skin.

Well, of course, the chief porter realized that here was the explanation of the lightness of the loads after

Tso Tz'u had carried them for a while, and he told the storekeeper who told the steward who told the Duke that it was all Tso Tz'u's doing. And in no time at all the Duke had Tso Tz'u arrested and thrown into a tiny cell in the state prison.

'He has fed enough at my expense,' said the Duke. 'There is no need to feed him any more.'

And so Tso Tz'u was left to languish and die, and the Duke forgot all about him. He was reminded of the incident only after a year had passed, when one of his earlier guests paid a new visit and asked what punishment had befallen the mischievous Taoist.

'To be sure,' said the Duke, 'it's time his body was taken out and burned.' And he gave orders that this should be done.

But when they lifted the bars and opened the door of the cell, what should meet their eyes but the sight of Tso Tz'u, rapt in meditation, but as ruddy-cheeked and firmly fleshed as ever he had been!

They hardly dared to report their finding to the Duke. But he was not a mean-minded man, and he realized that this was no common malefactor he was dealing with. He gave orders for Tso Tz'u's release and invited him to a dinner. At the end of the dinner Tso Tz'u announced that he was going away, and suggested that the Duke and he should drink a parting cup together. He took a cup of wine and asked for a pin. With the pin he drew a line across the middle of the wine. The wine split down the middle into two portions, which remained separate. Tso Tz'u drank off his portion and handed the Duke the cup, one side of which was still full of wine. But the Duke feared some fresh mischief and would not drink. Tso Tz'u then cheerfully drank up the rest of the wine, and threw up the empty goblet. It did not fall at once, but soared and hovered like a bird below the ceiling of the hall. Everyone watched in astonishment: and when they looked down again, no sign of Tso Tz'u was to be seen.

The tale of these wonders began to spread among the people of Wei, and the Duke grew uneasy at the thought of fresh mishaps that might occur. He regretted his leniency in releasing Tso Tz'u, and gave orders for him to be re-arrested. Soldiers scoured the countryside in search of him. At last a party of troops came across him on a hill-side. When Tso Tz'u saw them coming he ran off up the hill. Just above him a flock of sheep were grazing. The soldiers watched him run and started in pursuit. But suddenly, there was no old Taoist to be seen.

'Vanished into thin air,' said the captain of the soldiers in disgust. With pursed lips he surveyed the drifting sheep. Then inspiration seized him. He strode over to a rock on which lay, dozing in the sun, the shepherd of the flock. This man scrambled to his feet as the captain addressed him.

'How many sheep are in your flock?'

'Sixty-nine, sir.'

'Are you sure?' asked the captain.

'Sure? Of course I'm sure. There's Bosseye, and Shaggy, and Twirlyhorn, and Flatnose, and . . .'

'I don't want to know all the family!' shouted the captain. 'But we'll see whether you've still got sixty-nine—or seventy!'

And he ordered his men to count them. Unfortunately the sheep wouldn't stand still or form fours to be counted. There were sixteen men in the party, and the captain was given sixteen totals, between forty-two and ninety-three. In desperation he appealed to the shepherd, who was looking on with a grin like a split watermelon. The shepherd gave one glance at the flock, then started in surprise.

'You're right, Captain,' he said. 'There's seventy there. Now who's this come along all uninvited?'

'I know very well who it is,' said the captain. He turned to address the flock. 'Mr Taoist,' he said, 'I do urge you to come along with me. The Duke, my master, is waiting to welcome you.'

There in the middle of the flock, one of the sheep knelt down on the grass and bleated, 'Can I really believe that?'

'That's him,' said the soldiers. They were just making after him—when every sheep in the flock knelt down and bleated, 'Can I really believe that?'

Not wishing to arrest the whole flock, and not at all sure that he would have Tso Tz'u even then, the captain gave up his hopeless task and led his men away.

The Inn of Donkeys

THERE was once a merchant named Chao who did a great deal of travelling about the country. He knew all the best inns, but one day, following an unfamiliar route to the Eastern Capital, he found himself in a district which was new to him. Night was drawing on, and he asked some farmers in a field whether there was a good inn in the vicinity.

'Travellers in these parts always stay at the Wooden Bridge Inn, just over the hill there,' he was told. 'That's where they all buy their donkeys.'

'Have they good donkeys there?' asked Chao. 'This ancient creature which is carrying me is getting past it; I could do with trading him for a new one.'

'Best donkeys in the province,' came the reply. 'It's a widow keeps the inn, her name's Third Lady, and she's very rich, so she always sells her donkeys very cheap and doesn't have to worry.'

'Where do her donkeys come from?' asked Chao.

The farmers' faces took on a vague look. 'Now that we can't tell you,' they said. 'Better ask her—we don't know.'

Reflecting that country people seldom knew or cared what went on more than a mile away from their own farms, Chao continued his journey over the hill. And he came as directed to an inn, neat and trim and inviting in appearance, with comfortable-looking benches outside set among flowering shrubs, and a large signboard proclaiming it the Wooden Bridge Inn. He dismounted from the donkey he was riding and went inside, where he found Third Lady serving wine to some six or seven guests.

Third Lady proved to be a pleasant, cheerful woman of thirty or so. She gave him a polite welcome, then said, 'Perhaps you would care to stable your donkey and your baggage-donkey round at the back. I'm sorry I haven't a groom to look after them for you, but you see I don't keep any servants.'

As Chao led his donkeys to the stable he said to himself, 'No wonder she is rich if she has no servants' wages to pay. But what a capable woman she must be to run a smart inn like this all by herself. It's only surprising that someone hasn't snapped her up for a wife—but then, perhaps she's one of these loyal widows who refuse to remarry.'

He joined the other guests, all travelling merchants like himself, just in time to be served a most excellent dinner of tender slices of chicken, fresh fish from the river and finest white rice. Then while the widow washed the dishes the guests made free with the jars of wine she had placed on the tables. Chao munched sweetmeats and exchanged stories with the rest, but he was not a drinking man and left the wine alone.

Soon the time came to retire. The lamps were put out and the tipsy guests were soon sound asleep. Chao had been given a clean and comfortable bed next to a partition of rush matting, on the other side of which was Third Lady's own room. He lay for a while drowsily reflecting on the widow's extraordinary ability. The gentle snores of his fellow-guests were gradually lulling him away when suddenly a louder noise startled him into wakefulness. It was the sound of some heavy object rumbling across the ground, and it came from the other side of the partition, where the widow slept.

Chao's first thought was that thieves had broken in and were doing her harm. He sat up, keeping very quiet, and found a gap in the matting which he could peer through. With relief he saw that the widow was alone. The noise had been made by a heavy trunk which she had dragged into the middle of the room, where she now knelt. Then Chao gave a start. From the trunk the widow had taken the little wooden figure of a man, about a hand's breadth in height. This she placed on the earth floor of the room, while Chao continued to watch in lively curiosity. Next she took out from the trunk a wooden ox, and then a plough, both made to the same scale as the little wooden man.

Third Lady hitched up the tiny plough behind the ox and set the man behind the plough. Then she sprinkled water over them. Under Chao's astonished eyes the little team began to move, and in no time at all had neatly ploughed up the floor of the room. Third Lady now placed a tiny basket of seeds in the hand of the little wooden man, who proceeded to sow the field he had ploughed. Hardly had the seed touched the soil when up came the green shoots of wheat. And now the ears were full, the crop was ready. Third Lady gathered in her grain and knelt in the corner of her room where she threshed, winnowed and ground the ears into flour. With the flour she made cakes, which she put into the oven and baked. When she had finished her work she

went peacefully to bed and slept; and Chao on his side of the partition did the same.

When morning came the guests rose and washed and assembled for breakfast, which took the form of freshly baked wheaten cakes. But Chao had no intention of eating the one served to him. He hid it in his sleeve and left the room. Uneasy at heart after what he had seen, he collected his donkeys from the stable and prepared to leave. Before going he peeped in through the window at the other guests as they breakfasted. As he watched he saw them, one after another in swift succession, slump from their benches to roll on the floor. Their clothes changed into coarse hair, they grew tails and long hairy ears, and soon the inn was filled with the braying of half a dozen donkeys. Third Lady came in with a stick and herded them all out towards the stables at the back.

Chao urged his beast to a fast trot and hurried away from the scene. When he reached the Eastern Capital he said not a word to anyone of what had taken place, but arranged his affairs there and prepared to return home. One of his preparations was to buy some wheaten cakes of the same size and shape as Third Lady made, and in with these he put the one he had saved from breakfast at the Wooden Bridge Inn.

He stayed at the inn again on his way home. This time he was the only guest. Third Lady gave him the same polite welcome and cooked him an excellent dinner. Afterwards, since they were alone in the inn, she sat chatting with him, but failed to persuade him to drink the wine she had set out.

In due course each retired for the night. As he had expected, Chao had not been lying long on his bed when he heard the sound of the trunk being dragged out and opened. But this time, smiling to himself, he turned over and went to sleep.

In the morning the widow brought him tea and wheaten cakes for breakfast. But he had already set out on the table the cakes he had brought with him, and he said to her, 'I bought these in the Eastern Capital. They are specially good—won't you try one? Then you can keep your own cakes for tonight's guests.'

Third Lady was a little crestfallen but made no objection. Chao gave her the one cake of her own baking which he had carefully saved for this moment. She did not recognize it as her own, but thanked him for it and began to eat. No sooner had she taken three or four bites than she rolled on the floor and turned into a donkey.

Delighted with the success of his scheme, Chao prodded the donkey to its feet and examined it. It was big and strong, and a willing-looking beast, far to be preferred to his own jaded mount. Before riding away on its back, Chao rifled Third Lady's trunk for the little wooden puppets. He hitched up the team and sprinkled them with water, but could not make them work. So he burned them to make sure that they raised no more of their poisonous crops.

Chao's new donkey gave him splendid service for four years, until a day came when he was riding it down a street in the city of Changan. An old man coming towards him stopped and peered at the donkey, then clapped his hands and cried with glee, 'Surely it must be the widow Third Lady of the Wooden Bridge Inn? I'm delighted to see you so well—but how you have changed!'

Then the old man addressed the merchant Chao: 'She did a great deal of harm, and I must congratulate you on trapping her the way you did. But you have punished her enough, riding and beating her these past four years. Please let her go now.'

Chao dismounted from the donkey, and the old man drew a knife from his sleeve and slit open the beast's belly. Out from inside it jumped the widow of the Wooden Bridge Inn. She hid her face in her hands and ran away, and the merchant Chao neither saw nor heard of Third Lady ever again.

The Pavilion of Peril

LONG before universities were invented, students used to seek to advance their learning by 'wandering with lute and sword' through the whole wide realm of China. The peasant mending his dyke would not be too surprised to see, passing along the road above him, a young man of fine bearing, sword at hip and bag of books over his shoulder. Perhaps a lad would follow behind with the lute, for music was not the least important of his master's studies. Or perhaps the student, too poor to engage a servant, might carry it himself, for he would carry little else. He might be on his way to visit a famous battleground of ancient times, or to search out some poet who lived as a recluse in the near-by hills.

Such a young student was once making his way along a dusty road to the south of Anyang. The light was fading, and his hopes were set on finding some inn, or perhaps a monastery, where he could find shelter for the night. At last he came upon a little cluster of peasants' huts. Outside one of them an elderly woman was sitting, pasting together layer upon layer of cloth to make the sole of a shoe.

'Tell me, Mother,' the student addressed her pleasantly, 'can I find an inn hereabouts for a night's lodging?'

'Not before Ten Mile Inn, young master,' came the reply.

'And how far is that?'

'Ten miles from Anyang, and six miles along this road from here.'

The student gazed along the road which stretched ahead. It would be dark long before he reached the inn, and the prospect of six more dusty miles was not a pleasing one. But as he gazed he noticed, to the left of the road a bowshot from where he stood, a pavilion of a kind that many districts provided as a resting-place for travellers. It was in a sad state of disrepair, and looked as though it had long been out of use; still, it would protect him from the cold night wind, and he told the old woman he would rest there rather than go on. To his astonishment, a look of terror filled the old woman's eyes. Quickly she dropped her glance, then turned as if for help to a younger woman, no doubt her daughter, who had just emerged from the house. By this time also the menfolk were returning from the fields, and one of these, thickset and with a 'no-nonsense' look about him, asked what was afoot.

'The young gentleman wants to stay the night in the pavilion.'

'You can't sleep there,' said the man bluntly.

The student was annoyed by such rudeness, and wished to know by whose orders he should be stopped from using the pavilion.

'Nobody sleeps there,' said the man. Again the student questioned him, his temper rising, and a fierce argument seemed about to develop, but was halted by the raised hand of the old woman. She looked fearfully to left and right, then spoke in a hoarse whisper that still seemed too loud for her liking.

'Three men have slept in the pavilion in these three years past,' she said. 'Of the three, not one has lived to see the morning.'

'What tale of bandits——' the student began.

'Not bandits,' croaked the old hag. 'Not men at all——'

The student interrupted her. He knew better than to laugh at an old wives' tale, but still he answered her firmly and without fear. 'Then if it is spirits,' he said, 'it is time the spirits were disposed of. Come, let me share your evening rice, and then I shall trouble you no further.'

All muttered and shook their heads, but the evening meal passed without further argument. When he had finished the simple food the student gave a few cash to the woman, took up his things and wished the family a peaceful night. No one responded, but all followed his departure with the gravest looks.

The paint was peeling from the balustrade round the pavilion, and weeds pushed up through cracks in the stone steps. Inside all was darkness and the musty smell of decay. The student took from his bag a lamp and a book. He lit the lamp, seated himself bolt upright on the edge of a platform covered with filthy matting, and began to read.

Hours passed. Tired as he was, the student kept his attention on his book. No sound but that of the wind, rattling now and then a loose tile on the roof.

At last, footsteps on the road outside. From where he sat the student could see through a trellis the approach to the pavilion, faintly lighted by the stars. A man stood in the path, a man dressed all in black. In a deep voice the man called the master of the pavilion. A second voice, from within the room in which the student sat, answered, 'Here!'

'Who is in the pavilion?' asked the man in black.

'A student,' replied the master. 'He is reading and has not yet gone to sleep.'

The student watched the man in black turn and walk away; then he returned to his book. But before long, there was a fresh sound of footsteps. A man wearing a red hat stood before the pavilion and called the master.

'Here!' came the voice from the student's room.

'Who is in the pavilion?' asked the man in the red hat.

'A student,' replied the master. 'He is reading and has not yet gone to sleep.'

Nothing more was said, and the man in the red hat also turned and walked away. The student waited, but there were no more visitors. He therefore rose, went out, and himself stood in the path. He called the master of the pavilion in exactly the way that the man in black and the man in the red hat had done. 'Here!' answered the master.

'Who is in the pavilion?' asked the student.

'A student,' replied the master. 'He is reading and has not yet gone to sleep.'

'Who is the man in black?'

'He is the Black Swine of the North.'

'Who is the man in the red hat?'

'He is the Red Cock of the West.'

'And who are you?'

'I am the Old Scorpion.'

The student was satisfied. He returned to his lamp and his book, which he continued to read through the rest of the long night. There were no further interruptions.

No sooner had light dawned outside than the thickset farmer of the previous evening came hurrying along the road, a stout stick in his hands. He slowed down as he drew near the pavilion. Cautiously he mounted the steps and pushed open the rickety door. Dread gave way to delight when his astonished eyes beheld the student, unharmed, still sitting bolt upright on the edge of the platform, calmly engrossed in his book. The farmer flung his arms about him, and was carrying him off in triumph to breakfast in his house. But the student demurred. 'First gather together your family and your neighbours,' he ordered.

When all were assembled, the student led them into the pavilion and told them to search. They pulled down a rotting screen in a corner to reveal a scorpion, as big as a drum, with a sting several feet long. The wicked sting darted towards the student, but he had his sword already in his hand and with one swift stroke severed tip from sting and head from body. The little crowd gasped with relief. The student turned to them.

'Where do you keep a black pig?' he asked.

'In a shed to the north of the pavilion,' he was told. They led him to the place, and there he found a black pig, bigger than any he had ever seen. As he approached it raised its head, and there was no mistaking the evil gleam in its eyes. Again the student's sword rose and fell, and the black pig's head rolled at the feet of the astonished villagers.

'And where do you keep a red cockerel?' he asked.

'In a shed to the west of the pavilion,' he was told. They led him to the place, and there he found a cock with a great red comb and wicked-looking claws. For the third time the student wielded his sword, which he then wiped clean and replaced in its sheath at his hip.

From that day on, no traveller ever again had his peace disturbed in the pavilion south of Anyang.

The Rainmakers

In the ordinary course of events, we mortals merely have to accept whatever weather is sent to us. Nor do we have much warning in advance of what is coming. But now and again someone has chanced upon the rainmakers at work, and once upon a time one man even assisted them, though the result was not altogether a happy one.

It was by a chance encounter that Li Yung, a farmer, was able to save his wheat harvest. It came about when he was sent on an official errand to a town some distance away. Returning, he spent the night all alone in an old temple. In the middle of the night he was roused from his sleep by a loud banging on the stout temple door.

'Open up!' cried a voice from outside. 'We've come to borrow the thunder-cart.'

'Who wants it?' came the reply from within the temple.

'The King of Chieh-hsiu.'

Now Chieh-hsiu was the name of the district in which

Li Yung's farm lay, and so he listened with particular care to what followed.

'What does he want it for?' came again from within the temple.

'He wishes to take in the wheat harvest,' those outside replied.

There was a long pause, while Li Yung trembled to think that his wheat harvest was in danger.

Then the voice from inside spoke up again, 'Our King says the cart is in use just at the moment, and you'll have to wait a while.'

This time there was a longer pause. It was ended at last by a loud rumbling noise which came from outside. Li Yung looked out and beheld five or six spirit-men carrying candles and escorting a large cart, which they led round to the messenger from Chieh-hsiu. The rear of the procession was brought up by a number of men carrying a gigantic banner composed of separate streamers, each of which sparkled brightly. They presented the banner before the messenger, who counted aloud the number of streamers. There were eighteen. He nodded his head in satisfaction, and led the procession away.

As soon as the noise had died away Li Yung rushed out into the black of night and somehow or other found his way back to Chieh-hsiu. Though it was not yet dawn he roused every one of his neighbours by hammering at the doors and yelling, 'The thunder-spirits are coming to steal our wheat! Come quickly and gather in your harvest!'

But the neighbours only laughed at him and went back to bed. Li Yung dashed off to his field, where by working frantically he managed to reap and gather in his wheat. By the time he had finished it was broad daylight. He took his wife and children up on to a low rise overlooking the village, and watched the sky anxiously. They had only been there a few minutes when a cloud like thick black smoke rose up from behind the mountains. In no time at all it covered the sky. Torrential rain began to fall, thunder pealed and rolled, and Li Yung counted eighteen distinct flashes of lightning.

When the storm had passed the crops were ruined. The people of Chieh-hsiu were angry with Li Yung. The story spread that he had been practising black magic to enrich himself at the expense of his neighbours —they forgot how they had ignored his warning and laughed at him. So indignant did they become that in the end they took him to court. But there he told the whole story, and the magistrate, realizing that it was no more than a lucky chance which had shown Li Yung the rainmakers at work, pronounced him innocent and released him.

The man who once helped to bring the rain was also named Li, but he was no poor farmer but a warrior, Li Ching, who later became Duke of Wei. As a young man he was very fond of hunting. He used to go each year to the Ling mountains, where he would stay in a little village in a valley. The old men of the village thought a great deal of him and treated him most hospitably, and as the years went by his friendship with them deepened.

Towards the end of one long day of sport, just as he was thinking of returning to the village for the night, he caught sight of a fine stag above him on the hill-side. He spurred on his horse and gave chase. The stag made for a high ridge and sped along it like the wind. Li Ching, his blood up, followed; but his horse was tiring. The ridge sloped gently down into a wide, wooded valley. He failed to catch up with the stag before it reached the edge of the trees, and there he lost it. Out on the hill-side there had still been some light, but here, among the thickets, it was already dark. Li Ching soon realized that in trying to find the way out of the wood he was in reality plunging deeper and deeper into it. He was hopelessly lost.

He dismounted and looked about him: it was a frightening place, the grey wraiths of tree-trunks seeming to move and glide in the darkness, the silence broken only by the occasional mournful howl of a monkey. He led his horse forward, not knowing even what he was hoping to find.

The last thing he expected was a human habitation in the midst of this desolate forest. Yet in the distance across a glade a point of warm light glowed. He crossed the glade towards it. White in the gloom rose high walls, surrounding a large house. The great door was lacquered red and shone in the light of a bronze lantern. Li Ching hammered on the door with his fists. For a long time there was no response, but at last the door swung open and a servant appeared.

'I was hunting on the hill-side and lost my way in these woods,' said Li Ching. 'Can you give me shelter here for the night?'

'My masters are both away,' said the servant. 'There is only the old lady at home. I don't think she will allow you to stay here.'

'Please give her my respects and ask her if it is possible,' Li Ching requested.

The servant returned with the reply, 'My mistress was at first unwilling, but since you are lost and it is dark she will receive you. Please follow me.'

He led Li Ching into a hall, where a maid announced the entrance of the mistress of the house. She was about fifty years of age, dressed in plain but elegant style, and evidently a person of breeding. Li bowed to her, and she returned his greeting.

'Really it is not fitting that I should entertain you here when my sons are not at home,' she said. 'But I do not think you would be able to find your way out of the wood in the darkness.' Then she added, 'My sons will be returning soon. They make rather a lot of noise when they come in, but please do not let it alarm you.'

She arranged for him to be given a meal, which was sumptuous, but consisted almost entirely of various kinds of fish. Then servants brought in bedding, sheets and pillows, all soft and clean and pleasantly scented. When they went out again they locked the door on the outside.

Tired though he was, Li Ching was unable to go to sleep at once. He found himself wondering who these people could be, living in a mansion in the heart of a desolate forest. Why should the sons make such a noise when they came in—and where would they come from, so late in the night? Why had his door been locked on the outside? And something else puzzled him: why did they seem to eat nothing but fish?

He was still sitting on the edge of his bed trying to think of an answer when midnight came, and with it a loud banging on the gate.

A voice roared from outside the gate, 'Command of Heaven: the eldest master is to deliver rain over a radius of seven hundred miles. The rain is to cease at dawn, and there is to be neither omission nor undue violence.'

Li Ching next heard the lady of the house say, 'What am I to do? Neither of my sons is at home, and there is no time to send for them. None of the servants can undertake such a task, and yet I shall be punished if the rain is not delivered. Oh, what am I to do?'

Then a servant said, 'Would it not be possible to request the help of our guest? He seems to be above the common run of men.'

The lady expressed her delight at the suggestion. She knocked at Li Ching's door and asked, 'Are you awake, sir? I should like a word with you.'

Li Ching rose and went out to her.

'I must tell you,' began the lady, 'that this is no house of mortals, but a dragon palace.' ('So that is why they eat only fish,' thought Li Ching.) 'As you well know, we dragons are responsible for the bringing of rain. I have just received an order for delivery. But my elder son is attending a wedding in the Eastern Sea, and my younger son is escorting his sister home. Both of them are thousands of miles from here, it would be impossible to summon them in time. I wonder if you would help us in this emergency? It would not take you very long.'

'I'm afraid I'm only an ordinary human being,' said Li Ching. 'I am not a bad horseman, but I have had no experience of riding the clouds. However, I should like to be able to help you in return for your hospitality. If you will show me what I have to do, I will try my best.'

A piebald horse was led forward. The lady called for the rain-jar. To Li Ching's surprise this proved to be quite a tiny jar, which the lady fastened to the piebald's saddle.

'There is nothing difficult about it if you will simply do as I say,' she said to him. 'You will not need to use reins or whip. Simply let the horse have his head. But whenever he stops and neighs, there you must shake one drop of rain from the jar on to his mane. Be very careful that you use no more than one drop each time.'

And so Li Ching mounted and set out on his mission. The piebald rose into the air, higher and higher. At first Li Ching dug in his thighs from fear of falling, but soon he realized that the horse was as steady as a rock. It was hard to believe that he was travelling above the clouds—yet the wind flew past him like a flight of arrows, and beneath him lightning played and thunder rolled. The horse stopped and neighed, and Li Ching followed his instructions, dipped a finger in the jar and shook one drop of water on to the piebald's mane. This happened several times, and each time when he had done it the clouds opened beneath him.

Eventually the horse stopped above a place which Li Ching recognized. It was the mountain village in which he had been staying while he hunted.

'The people of this village have been very good to me,' he said to himself. 'I have received a great deal of hospitality from them, and I have never found any way of repaying their kindness. But these last few weeks have been a time of drought, the crops are parched, and here am

I sitting with the rain in my hands. I will give them a good downpour to water their crops.'

And so this time he shook not one drop but twenty drops of rain on to the piebald's mane before moving on. This was in fact the last of the stopping-places. The horse made straight for the house in the woods, and there Li Ching found the lady waiting for him. To his astonishment she was weeping. As he dismounted she cried out to him, 'Why did you not do as I said? I asked you to shake one drop of water on to the horse's mane, and you used twenty! The reason I said one drop is that one drop from the rain-jar causes one foot of rain on earth. And there is that poor valley, in the middle of the night, flooded out by twenty feet of rain! None of the inhabitants can have escaped!'

Her tears burst forth again. She turned her back to him and removed a shawl from round her shoulders, revealing angry red weals across them. 'I have been punished with eighty strokes of the rod,' she told him. 'My sons will be punished also for having neglected their duty.'

Li Ching was filled with horror by the consequences of his act of kindness. Noting his remorse, the lady went on, 'Of course, you are only a mortal and could not be expected to understand the art of rainmaking. I must not be angry with you. But I am afraid of what my sons will say if they find you here. You had better leave at once.'

She clapped her hands, and two young slave-girls entered, one through a doorway in the east wall of the hall, one through a doorway in the west wall. Both were outstandingly beautiful. 'Here in the depths of the forest,' said the lady, 'we have nothing with which to reward you for your labours on our behalf. All I can offer you is these girls. You may take one or both: but be very careful in your choice.'

Li Ching looked at the two girls. He found that one was smiling at him; the other was glaring at him with a hostile expression. He said to himself, 'I am a hunter and fond of fighting. If I choose the girl who is smiling at me the lady will think I am chicken-hearted.' And to the lady he said, 'I would not presume to accept both. I will take the girl who came through the doorway in the west wall, and who now is glaring at me so fiercely.'

'So that is your choice,' said the lady with a wise smile. She led the girl up to him, and then, before she sent them away, she put into Li Ching's hand a bag full of large, flawless pearls.

Li Ching looked back when he had ridden with his bride a few paces from the house, but there was no longer any sign of it. Dawn came, and he found his way with ease back to the mountain village. As he had feared, it was a sight to chill the flesh. Only the roofs of the huts showed above the muddy waters of the lake which now filled the valley. But as it had happened, he had been clumsy and slow in taking the twenty drops from the rain-jar. The people of the village, warned by the storm of thunder and lightning, had had time to escape up the hill-side, where Li now found them bewailing the destruction of their homes.

He comforted them as best he could, then rode off to sell the pearls the lady had given him. With the fortune they brought him he was able to feed the villagers until their land was reclaimed, and build a fine new village for them where the old one had stood.

In later life, as Duke of Wei, Li Ching won undying fame as a general. This was because he had chosen the slave-girl who came through the doorway in the west wall, the girl who had scowled at him. It is often said,

'Ministers of state come from east of the mountains, generals come from the west.' If Li Ching had chosen the smiling girl who came through the doorway in the east wall, he would have become a minister of state; if he had presumed to accept both the girls, then he would have been both general and minister at the same time.

THE REVOLT OF THE DEMONS

I

Aunt Piety

ON a snow-covered plain in northern China, under the faint starlight of a night many centuries ago, the only living things to be seen were an old peasant woman and her young and comely daughter. Aunt Piety was the old woman's name, and there was nothing in her appearance to indicate that her proper home was no peasant cottage but a fox's den. Such is the power of the fox, which of all magical creatures has the greatest gift for assuming human form. For Aunt Piety was indeed a fox, and so was her pretty daughter by her side, and so was her son, Blackfoot, whom they had left behind.

It was by a moment's carelessness in the use of his magical gifts that Blackfoot had landed himself in trouble. Not many weeks previously, his covetous eye had lighted on the pretty face of a farmer's wife. Using his spells on a night of full moon, Blackfoot turned his red fur and bushy tail into the form of a young gallant.

But in his courtship of the farmer's wife he had carelessly aroused the husband's suspicions. The honest farmer took his bow and followed a trail of tell-tale footprints in the dew. Across the fields he traced them, and emerged into a moonlit glade at precisely the moment when Blackfoot, head raised to the full moon, was returning to his fox-shape. Quick as a flash the farmer took aim. A howl broke the silence of the glade, and Blackfoot dragged his wounded hind leg in anguish back to the den.

Aunt Piety knew very well that her own efforts would not suffice to mend her son's leg. Her only course was herself to assume human form and go for help. Dressed as an old beggar-woman she sought out the one doctor, Yen, whose fabulous skill could meet this complicated case.

Yen's sharp eyes were not deceived by Aunt Piety's disguise. 'I take it you have come to see me about your son,' he said with a slight smile. 'Perhaps there is something wrong with his *tail*?'

For all his witticisms the doctor gave Aunt Piety more than a mere prescription for her son's injured leg. He gave her forewarning of the terrible fate that lay in store for her beautiful daughter. This could only be averted, said he, if the whole family were to leave their den and cultivate with might and main the *Tao*, the True Way.

And this was why Aunt Piety now trudged with her daughter through the snow. Blackfoot's leg, mended but left for ever lame, had proved such a handicap that they had left him resting in a temple. When they had found the haven they sought they would summon him, so that then the three of them could cultivate the Way together.

The night was cold. Even in their human shape the

two foxes had little to fear from this. But they were weary from walking, and nudged each other with relief when they saw before them a deep thicket in which they could take their rest. It was not more than a mile or two away. But as they approached, suddenly a great wind arose, a black wind which blotted out all before their eyes. They clung to each other in terror. The wind passed, Aunt Piety opened her eyes, and there before her two armed warriors bowed in salutation.

'The Queen of Heaven requests the presence of Aunt Piety,' they said.

'The Queen of Heaven?' asked Aunt Piety. 'What other name is she known by?'

'She was known as the Empress Wu in the time of the Tang Dynasty,' replied the warriors.

A shiver went through Aunt Piety at these words. The Empress Wu, the woman who had usurped the Dragon Throne to rule the Tang Empire with a rod of iron! But the Empress had been dead for hundreds of years. And how could she, Aunt Piety, have come to the notice of so august a figure?

Such questions hummed through her mind as she was led by the warriors through an overgrown path between great looming trees into a courtyard where palace women waited to take her into the presence of the Queen of Heaven. Aunt Piety prostrated herself, but the Queen was all friendliness. She seemed to welcome a sympathetic listener as she spoke of her own sorrows. In her time on earth she had tried to rule with virtue, she claimed, but had been betrayed by unscrupulous ministers. Before she had been dead many centuries her tomb had been rifled.

'It was in a time of chaos and revolt,' she sighed. 'Nothing was left, I am ashamed to appear before you with not a bracelet to my name.'

Aunt Piety looked up and saw with dismay that the Queen indeed wore no ornaments of any kind.

'How was it that your spirit had no power to stop these ruffians?' she asked.

'It was decreed by fate, I was helpless,' sighed the Queen. Then her face brightened. 'But now the fates have been kinder to me. On earth I most desired to be no mere woman but a man. I played the part of a man. And now my wish is to be fulfilled. I am to be reborn, to rule again, but this time at last as a man. And it is your daughter, Aunt, who is destined to be my consort.'

Aunt Piety looked up in amazement as the Queen continued, 'You and I will meet in Peichou twenty-eight years from now. You can help me greatly in the meantime by diligence in your study of the magic arts.'

'It is in search of the True Way that I am journeying,' said Aunt Piety. 'Where should I find these magic arts to study?'

'Listen carefully,' said the Queen.

'*The Aspen will detain,*
The Egg will all explain.
Seek not: yourself be sought;
All other search is vain.'

Aunt Piety nodded her head, committing these precious words to memory. The Queen added a final warning. 'What I have told you belongs to the secrets of Heaven, which must never be divulged. Above all, do not let this reach the ears of my arch-enemy, the Octogenarian.'

'Who is the Octogenarian?'

'Prince Yang of Han . . .' but the Queen was interrupted by a palace maiden, who rushed in, forgetting all ceremony in her terror.

'Prince Yang is here with a hundred thousand warriors,' she screamed. The Queen's face turned grey. She whirled round and fled.

'Wait for me!' yelled Aunt Piety. Jumping forward in her anxiety she kicked over a stool. There was a deafening crash—and Aunt Piety found herself, bathed in cold sweat, shivering in the dark depths of an empty grave. Queen, palace, all were gone: she was alone in the silent wood.

The really alarming thing was the disappearance of her daughter. This was the proof that what Aunt Piety had been through was more than just a dream. With some difficulty she clambered out of the violated tomb. A yard away from it she stumbled over a large stone. Clawing away the brambles and weeds that twined over its face she was able to make out a row of engraved characters. The stone had once stood upright, and bore the name of the Empress Wu.

'The Aspen will detain,
The Egg will all explain.
Seek not: yourself be sought;
All other search is vain.'

The Empress' words repeated themselves unfalteringly in Aunt Piety's mind. The Aspen must be the first object of her search now: but where to find it? Her son she had left days before in the temple. Her daughter now was gone: was this in fulfilment of the learned doctor's prediction of an evil fate in store for her? Aunt Piety shivered. But then she remembered again the Queen of Heaven's words, that after many years the girl was to be her consort in Peichou. Perhaps she had gone now to prepare for this event.

All must be determined by fate. For the time being

Aunt Piety could do no more than go forward to the Great Hua Mountain, where she had originally planned to cultivate the True Way.

After many days Aunt Piety reached and climbed the Great Hua Mountain. She found there, gathered in innumerable temples and shrines, hidden away in secluded hermitages, or merely wandering with bowl and staff from place to place on the Mountain, a great concourse of priests and nuns, old and young, but all poor, ragged and in constant search of alms. From time to time a layman, an official, a merchant, or more often a wealthy woman from the city would ride or be carried up the Mountain to worship. It was on these occasions that Aunt Piety realized to the full how dense was the Mountain's population. For whenever a visitor appeared who gave promise of some substance he or she would become at once the focus of a swarming crowd of mendicants.

One day Aunt Piety heard the shrill voice of a wealthy merchant's wife ordering away a crowd of begging nuns.

'I have brought no money with me today. Why do you come to me for alms? Do you think my husband is Old Buddha Yang, that I should give to all and sundry? Why don't you go to him and that pious wife of his if you want a good meal?' She went on in this strain for some time.

Now, the meaning of the surname Yang is 'Aspen'; and the sound of it caused Aunt Piety to prick her vixen ears.

The Aspen will detain . . .

Eagerly she sought out an old Taoist woman whom she knew to be well acquainted with the vicinity. From her she learned that 'Old Buddha' was the nickname

of the Prefect Yang who lived in the near-by city, a name bestowed on him in honour of his practice of freely entertaining all persons of religion who chose to present themselves at his gate.

Without more ado Aunt Piety hurried down the Mountain and inquired her way to the Prefect Yang's residence in the city.

Reaching the gate she was surprised to find a notice inscribed in large characters. 'Feasts open to all on the first day of each quarter,' the notice ran. 'No alms given at other times.'

Aunt Piety sought out the gate-keeper and asked him the meaning of this extraordinary announcement.

'Well, as you probably know,' replied the gatekeeper, 'the master has always been very free in his giving, any time of day to anyone who called. But a month ago we had a real upset, and it put a stop to all that. A fine holy-looking nun wangled her way in and was entertained by the master for days on end. Then along comes this band of ruffianly priests at dead of night, out slips the nun and opens this very gate, and between them they make off with I don't know what in gold and silks and loot of all kinds. So you see the master's grown a little more cautious nowadays, and as for you, my good woman, you will have to wait till the first of the quarter if you want to fill your rice-bowl.'

But now she had found her 'Aspen', Aunt Piety was

in no mood to wait a single day. She kept the gate-keeper in conversation, and was beginning to work her way into his good graces when suddenly a procession emerged from the house. Flute-players and flag-carriers led the way for a sedan-chair carrying the richly gowned figure of the Prefect Yang himself. He was carried through the gateway and off down the street.

'Where is he going?' asked Aunt Piety.

'He'll be going to fetch the Sanskrit Scripture,' answered the gate-keeper.

'Can your master read Sanskrit, then?'

'Nobody can read this Scripture,' said the gate-keeper, 'but it is very precious. It was left by a deaf and dumb priest who died at the shrine by the city wall. It is all written in gold in Indian characters, and the master's had a special box made for it.'

This was Aunt Piety's opportunity. In her long fox-lifetime she had mingled with a wide variety of foxes, some of whom were very gifted, and one of the things she had picked up was a knowledge of Indian characters. She told the gate-keeper of this, though, of course, without mentioning the foxes she had studied under. The gate-keeper at first derided her claim, but began to be impressed by her tales of the holy saint, the Bodhisattva, who she said had been her tutor. At this the gate-keeper agreed to report her offer of help to the Prefect Yang.

'Very good,' said Aunt Piety. 'And tell him if he wants me at any time, all he has to do is face the south-east and call my name, Aunt Piety, three times.'

In due course the Prefect returned with his precious Scripture and the gate-keeper told him of the strange visitor. Yang was not inclined to take much notice, and decided to leave things for a day or two and see if she called again. But when he entered the house his wife hastened up to tell him of a wondrous experience:

'I was just in the garden looking at the pomegranate flowers, when I saw a beautiful coloured cloud coming up from the south-east. And there, right in the middle of the cloud, was a Bodhisattva all gorgeous in ornaments of gold and pearl, and with a beautiful serene face, and sitting on a white elephant. I knew it was a Bodhisattva coming to me in a vision. I bowed down to the ground but when I lifted my head to look again it had gone. I did not know what to think but then the gate-keeper told me about this Aunt Piety. It must have been when she came that the Bodhisattva came along with her.'

Of course, Aunt Piety was no mean performer in the art of changing herself into one thing or another. It was a matter of no great difficulty for her to conjure up a white elephant and ride for a few minutes on a coloured cloud. But the Prefect Yang, like his wife, was by now convinced that Aunt Piety was the holy woman she claimed to be. He turned to the south-east and called her name three times. At once a servant reported Aunt Piety's arrival at the gate. They welcomed her into the house and set before her the rarest of vegetable dishes. After the meal the Prefect Yang brought out from its box the priceless Sanskrit Scripture. Aunt Piety recognized it at once as the celebrated Heart Sutra, and to Yang's unbounded delight read out and interpreted the opening section of the text. Anxious to make the fullest use of such available services, the Prefect Yang invited her to stay as long as she wished in a retreat that stood ready and waiting in a secluded corner of the estate.

They made arrangements for Aunt Piety to be supplied each day with the choicest of meals, but she laughed at this.

'There is really no need for you to take such pains,' she said, 'it is no uncommon thing for me to fast ten years together.'

This was really an impossible claim, thought Yang and his wife, and they decided to try her out. For seven days no one went near her. On the eighth day Madam Yang paid a visit to the retreat, and was astonished to find Aunt Piety in the full bloom of health and vigour. Naturally, as a fox she was indifferent to human food, and laughed off Madam Yang's inquiries as to whether she was not desperately hungry. But these remarkable powers served further to convince the Yangs that their guest was truly a living Buddha, and from that time forward Aunt Piety was entertained with the utmost reverence in their house—fully content to be 'detained' by the 'Aspen'.

2

Eggborn

IN the hills of another part of China, a small pond lay just before the gate of a Buddhist monastery. The kindly old Abbot of this monastery, invited one day to conduct a service in the near-by town, reflected that his robe would first need to be washed, and went down to the pond with his bucket. It was then that he saw the egg which floated on the surface of the water. As the bucket was lowered into the water the egg seemed somehow to be drawn towards it, to fall with a 'plop' over the edge before the bucket was half full. At first the Abbot took it for an empty shell, but on picking it up he found it to be a large, whole egg, more like that of a goose than of a hen.

'Curious,' the Abbot said to himself. 'There is no one just here who keeps geese. Well, we'll see if it is fertile. If it isn't, the little novices can have it with their rice. If it is fertile, Old Chu's hen in the village might be able to hatch it out, and that will be a life saved.'

The Abbot held the egg up to the sunlight. A dark

mass within told him it was indeed a fertile egg. He sent it to Old Chu, and thought no more about it.

Just seven days later the Abbot was startled from his meditation by Old Chu himself, who rushed in with alarming news. The egg had hatched out—and now, at one end of the basket lay the mother hen, dead, and at the other a heap of empty egg-shells, whilst in the middle sat a little baby, six or seven inches long and perfectly formed.

Old Chu would have nothing to do with such a monster and insisted that the Abbot should take it away. The Abbot agreed that not a word of this should leak out to spread terror in the neighbourhood. Nothing but evil could come of such a birth, he was convinced. He carried the basket to a tiny secluded court of the monastery, where he dug a hole and buried the baby in the basket as its coffin. Leaving the garden he looked back, still trembling to think of what he had done. But even as he looked, the head of the child thrust up through the soil. In a rage of fear the Abbot struck at it with his spade. He missed, and the spade broke into fragments, while the baby opened its eyes and gave him the sweetest smile.

Tears came into the Abbot's eyes.

'Poor creature,' he said. 'If only you had found re-birth in human fashion, what joy you would have brought to some fortunate couple. But you have taken the wrong road. It is not my fault, but I must send you back to try again. And next time try not to frighten people so much.'

With these words the Abbot again quickly covered the head with earth. This time he built a high mound over the tiny grave, and placed large rocks on the top. As he left the courtyard he shut and locked the gate, sure now that if the monster did not suffocate it would at any rate die of starvation.

For fifteen days no one went near the courtyard. At the end of this time the Abbot timidly unlocked the gate to see what might have taken place. As he had half feared, the rocks and earth of the mound were scattered in all directions. The grave was empty—and there, in a willow-tree, sat the child, naked, grinning, and by this time fully two feet high. At sight of the Abbot the child ran up to catch at his sleeve. It made no sound but its face beamed delight. The Abbot turned and fled. When he had calmed down again he began to doubt his own judgement, and decided to draw a prediction in the temple of the Goddess of Mercy.

The slips were tossed in the air, and from the scattered heap on the floor the Abbot drew out the Goddess's prediction. The slip selected bore the words 'Most Happy Augury'.

At last the Abbot could set his mind at rest. He rescued the child from the courtyard and gave it into the keeping of Old Dog Liu, a gentle servant of the monks who had lost his wife and had never had children of his own. The harmless falsehood was spread about that the child had been left at the gate by an impoverished peasant family. As time passed its rate of growth slowed so that the Abbot no longer feared that its head would one day crack the sky. The child learned to speak and began in every way to resemble an ordinary human child. The strange circumstances of its birth were, however, commemorated in the name, Eggborn, by which it came to be known. There were not a few of the monks who from time to time directed their malice against this 'child of an egg, reared by a dog'.

He had indeed certain habits which did not fit too well into the pattern of monastery life. Even as a small child he showed such relish for meat and for wine that it seemed unlikely he would ever be brought to forswear these things as the other monks did. By the age of fifteen he was well versed not only in the Buddhist Scriptures but also in the handling of a stout staff, the only object resembling a weapon of combat that was to be found in these monkish surroundings.

Unfortunate, too, was Eggborn's one failing, a hot temper. Time and again he would only with the greatest effort suppress the impulse to crack the skull of some sly monk who had insulted him. He often felt, indeed, that it was only his devotion to Old Dog Liu which prevented him from leaving the monastery to seek his fortune in the outside world. At length the time came when Old Dog Liu contracted a severe illness. For days and nights on end Eggborn, without once removing his clothes to take his rest, nursed the old man with the devotion of a true son; but his best efforts failed, and his grief knew no bounds when the old man died.

A strong tie was now loosened, and it was not long before Eggborn, driven to it by one last violent quarrel with his malicious fellows, seized his bowl and staff and walked out of the monastery gate for the last time. Months and years passed while he wandered through the length and breadth of the country. Begging his food wherever he went, never failing to pay his respects in any shrine he came to, Eggborn came to know the hills and ranges of China as a gardener might know the rocks in his courtyard. Often he would fall in with a monk or a Taoist priest engaged on the same unwearying pilgrimage, and they would journey together for a few days or weeks.

Eggborn was deeply conscious that his journeyings were in no way haphazard, but were carrying him towards the fulfilment of his destiny. So it was that he met with no obstacle, no hindrance to his cloudlike progress, until he reached the region of the Dream Cloud mountains. It was here one day as he crested a pass that he came face to face with the mist. This was no new thing to a traveller in the mountains; but he had never before seen a mist like this. Like a white-painted screen it blocked all view of what lay before him; he might be standing at the end of the world. He was roused from his perplexity by a call from behind, 'Turn back, turn back, you have missed your way!'

Eggborn turned to find a priest, a stranger, who beckoned earnestly to him.

'What place is this?' he asked at once.

The stranger took his arm and as they walked back down the pass began to explain: 'This is the region of the Cave of the White Cloud, the home of the White Monkey Spirit. He is the Guardian of the Text of Heaven, and the mist is his defence against the mortal world.'

'What is this Text of Heaven,' asked Eggborn, 'and how did it come to be placed in these hills of the mortal world?'

'That is a story known to few. The White Monkey Spirit, after long eras of self-cultivation, was given appointment by the Jade Emperor as "Guardian of the Texts of Heaven". For long he obeyed the injunction to keep his seals intact, and was all that a guardian should be. But perhaps a monkey never can be fully rid of his monkey-nature. The day came when all the spirits were invited to the birthday feast of the Queen Mother of the West—all except our White Monkey. Rather offended, and very bored, he disobeyed his orders by taking a peep at one of his Texts. It promised to be so full of profit that he decided to make it known to all his monkey tribe. He came down to earth and made a copy of the entire text which he left in their cave. But he had no sooner finished than the spirits returned from their feast.

The Jade Emperor was furious to learn that both Text and Guardian were missing. There was an arrest and a full-scale trial and the upshot of it was that a new Guardian was appointed whilst the White Monkey was condemned to watch for ever over the copy he had made, in the cave from which all other monkeys and all mortals are banned.'

'That is a very strange story,' said Eggborn. 'But you haven't explained the chief mystery. What is the Text about?'

'I believe,' the stranger replied, 'that the Text in the cave offers full instructions in one hundred and eight magical transformations.'

'Transformations!' exclaimed Eggborn. 'No wonder the Jade Emperor was so angry. And the mist, I suppose, sets an impassable barrier in the face of prying mortals?'

'Quite so,' said the stranger. 'For three hundred and fifty-nine days in the year the mist lies there, impenetrable. Only on one day in the year does it lift. On the fifth of the fifth month, for the two-hour period about midday, the Monkey Spirit ascends to Heaven to make his annual report. During these two hours the cave is defenceless. At the end of this time white smoke from an incense-burner announces the return of the Monkey, and the mist descends again.'

'Then no attempt has ever been made on the secret of the cave?'

'There was once a rash Taoist,' said the stranger. 'At the time the mist lifted he set out for the cave, but came back with a tale of a narrow bridge over a chasm so fearful that even in the clear light of day he dared not attempt to cross.'

They talked on for a while, then parted company. The stranger's story had roused a strange emotion in Eggborn, a feeling of certainty that he was the man to whom the cave would finally yield its secret. It was now the beginning of the fourth month; he had only a few weeks to wait. He gathered branches and creepers and built himself a rough hut in a fold of the hills near the pass. There he settled down to prepare himself for the great venture. But the days passed with agonizing slowness. He turned to wine to speed the time along. To his dismay this only served to increase his impatience. Finally one night, his brain reeling from the fumes of wine, he lurched out of the hut and battled his way into the mist. In less than a couple of miles the acrid mist in his throat and lungs joined with the vapours in his head to bring him to a standstill. He crawled miserably back to his hut, where he fretted away the days that remained of his waiting.

When Eggborn woke on the morning of the fifth day of the fifth month it seemed to him that the mist which filled the mountains was already less dense than usual. And indeed, just as he had hoped and prayed, as noon approached the mist cleared completely and the sky stretched blue and clear above the green hills. Eggborn put on a pair of strong hemp sandals, took a stout staff in his hand, and strode out eagerly on his mission. Even in the first mile the going became appreciably harder. There was no patch of ground which did not either sheer up or plunge down before him; and soon he was in the midst of dense thickets which were all but impenetrable. Still he found a path of a sort, a mere trace of a line which seemed to have been trodden for him. For several miles more the path led on, ending at last at the brink of a deep chasm. It was sickening to gaze down at the rocks which jutted out from foaming water far below like so many swords and spears. The chasm was some thirty feet across; but before Eggborn, as he stood there, an arch of rock formed a natural bridge. It was little more than a foot wide. One slip would mean certain death. Hardly stopping to think, Eggborn drew a deep breath and rushed across. Now before him a tunnel ran through living rock. Above the entrance to the tunnel he read the words, carved in the stone, 'Cave of the White Cloud'. He was there!

Passing through the tunnel Eggborn emerged into a valley of unearthly beauty, of smiling fields and gentle hills. Near at hand was a steep rock face. The view of the valley from the summit of this rock was obviously something not to be missed.

'Who cares about secret books, even the Books of Heaven,' Eggborn said to himself, 'with a landscape like this to feast his eyes on?'

He made his way to the rock face. At its foot stood an incense-burner of precious jade, the most beautiful thing Eggborn had ever seen. He examined it carefully and admiringly before beginning his climb. The view from the summit was indeed breathtaking. But before Eggborn's eyes had travelled half-way around the horizon, his nose was assailed by a strange perfume. Quickly he looked down to see white smoke rising in puffs from the incense-burner below.

It was the signal: the White Monkey Spirit, the Guardian of the Cave, was returning from his mission to Heaven. Panic filled Eggborn's heart, he came leaping down from his vantage-point, he raced out through the tunnel. Already wisps of mist were whitening the air. Eggborn rushed on to the stone bridge.

In his blind haste he slipped on the smooth stone, staggered, and only by a split second saved himself from hurtling to destruction on the jagged rocks below. Once across the bridge he paused for breath and regained something of his courage. The mist now was thickening fast, and Eggborn struck out manfully on the three or four miles which would remove him from its threat.

Back at his hut he reflected dismally on the failure of his attempt.

'What a fool,' he told himself, 'all that way and all that waiting just for an eyeful of scenery, and you still don't know whether there is any such thing as a Book of Heaven there or not.' There were fully three hundred and sixty days to pass before another opportunity would present itself. He groaned. Soon, however, his natural cheerfulness began to return. 'Well, at least I know the way now. There will be no difficulty next time—I crossed the bridge once and I can do it again. And next time the scenery can look after itself while I go straight to where the books are hidden and sling the lot of them over my shoulder. Then when I have got them back here I can choose which ways will be best to study.'

Thus encouraged, Eggborn decided to keep to his ambition. There was in any case no need to spend the year in moping about his hut. And so he left the hut and roamed the near-by country, begging his food, paying his respects at shrines and temples large and small, and now and then finding a useful service to perform for a fellow-creature in distress.

The fifth of the fifth month came round again at last, and this time found Eggborn better prepared than before. In the preceding week he had laid in a stock of provisions. And this time, mindful of the challenge of the chasm, he drank no wine but spent his days in meditation and prayer and the strengthening of his spirit for the ordeal. When the day dawned he set out early, braving the mist, and somehow or other fought his way through to reach the chasm at the precise moment the mist had cleared sufficiently to attempt the crossing.

Again he made his way successfully over the perilous stone span. Through the tunnel he went, straight to the spot where stood the jade incense-burner. Deliberately averting his eyes from the beauties of nature on every hand he sought for any such place as could hold the precious books he was after. His eyes lighted on a black fissure in the hill-side, straight across from the rock face he had climbed on the previous occasion. On closer investigation the fissure proved to be an entrance to a cave. Passing through, Eggborn found himself in a dim chamber perhaps five times as spacious as an ordinary room. From the far end of this a second, much smaller chamber opened. Here must be the hiding-place! Eggborn lowered his head and crept through into the smaller inner cave.

Now he was in a cosy little room whose walls, floor and ceiling were all of stone. The room was fully furnished with a bed of stone, a table and stools of stone, and a low writing-desk on which stood ink-tablet, water-jar and brushes, these also of stone. Eggborn fingered them, but each object was carved from the living rock and not one of them could he lift up. He explored every inch of the room, but there was no sign of so much as an account book, let alone any Book of Heaven. Beginning to despair he stooped and returned to the larger chamber of the cave. As he straightened up now his eyes took in for the first time the curious markings on the high, smooth rock wall before him. With a great shock of delight he realized the truth of the matter: that the Text of Heaven had the form of no ordinary book, written on paper and bound with thread, but was chiselled into the rock wall of the Cave of the White Monkey. He was face to face with the secret, sacred Text. Eggborn rubbed his eyes, fixed his gaze on the top right-hand corner of the wall and prepared to read. But precisely as he did so the scent of incense wafted through from the outer air: for the second time the attempt was foiled by the untimely return of the Monkey Guardian.

Without even a thought for the pity of it all Eggborn ran as fast as his legs would carry him out of the cave, through the tunnel, over the narrow bridge and the length of the path through the hills to his little hut. Back once more where he had started Eggborn surrendered himself to his disappointment. The whole year of waiting, a perilous journey, an exhausting search, the final discovery of the sacred Text itself, and all to no avail. He began to moan and then to weep. The Text, as he had seen, fully covered the broad high wall of the cave. Even if he were to wait yet another year, how should he possibly have the time to memorize it before the White Monkey returned? He began to sob aloud. And even if he were to take along brushes and paper, what hope had he of copying any useful portion in the brief time at his disposal? He cried and cried for three days and three nights.

At last he became aware of a voice calling outside, 'Who is this, in this place of quiet, that makes such a sorrowful din?'

Eggborn looked out to see an old man with a long white beard standing on the path. Somewhat cheered to find a confidant, he poured out the whole story of his woes. At the end of it the old man nodded his head. 'Yes,' he said, 'it is a very difficult position. As it happens, I too have visited this cave, and I am aware that the inscription is indeed of great length. However, you would have plenty of time to copy it if it were in your fate to do so.'

Eggborn looked at him in surprise, and the old man went on, 'It would be useless to hope for any brush to copy such a Text in the time allotted. But I learned once from a Taoist magician that the inscription would copy itself on to sheets of fine white paper placed over it by the right man. This man would have made a solemn vow to use his sacred knowledge only for good and not for evil. Then, if it were his destiny, the inscription would take; if it were not, the sheets of paper would remain blank.'

Eggborn began to thank the old man for this precious advice, but it was no use—he had already disappeared. He left Eggborn oblivious of past efforts and past disappointments. All that mattered now was to get somehow or other through the next three hundred and sixty days and make a fresh and final essay against the Guardian's treasure.

Once again Eggborn left his hut, left the hills, and wandered month after month through the countryside, begging his food, saying his prayers and performing deeds of service to those in need. As the fifth of the fifth month approached he made his way for the third time to the little hut in the hills. Again he fasted and purified himself, and this time he laid in a stock of one hundred sheets of finest white paper, which he carefully numbered in order.

The fifth day came, and before the mists began to lift from about the Guardian's cave, Eggborn was well on his way. He reached the narrow bridge. But then to his dismay the mists lifted, only to be replaced by pouring rain. The span of rock, already as smooth as glass, was now surely impassable! But determination showed Eggborn the way. Tying the bundle containing the paper securely across his shoulders, he lay down on the bridge and allowed his body to swing underneath it while his hands clasped tight across the glistening top. The under-surface of the bridge was not quite so

smooth, and by using his feet against it, hanging upside-down like a giant bat, Eggborn was able to make his way inch by inch across the terrible chasm. Once across he ran through the tunnel straight to the smokeless incense-burner. There he knelt and made a solemn vow: 'I, Eggborn the priest, do hereby solemnly swear that should the secrets of Heaven be vouchsafed to me, I will use the knowledge I thus shall gain to do good in the world and no evil. All my acts shall be in accord with, and none against, the Will of Heaven. Should I transgress this vow, may Heaven punish and earth destroy me.' He rose, turned and ran up the hill-side, into the cave.

Without pausing for breath, he drew the sheets of paper from his bundle and began carefully to place them, one after another, against the left-hand wall of the cave. It took only thirteen sheets to cover the inscription here. But that on the right-hand wall was longer, and after twenty-four sheets Eggborn still had not reached the end of it. As he was taking out the next blank sheet from his bundle the now familiar smell of the Guardian's incense reached him. Sad that he could not complete the task, yet delighted with what was already in his possession, Eggborn carefully packed the used sheets of paper, left the blank sheets where they lay and hurried, his heart singing within him, back to his hut.

Now was the time of testing. He had made the vow to use the secrets of the Text only for good. If it were his destiny to learn the secrets, the sheets of paper in his bundle would be covered with impressions of the Texts; if not, they would be blank. For a moment Eggborn hovered between his eagerness to open the bundle and his fear of what he might find within. Then, his heart pounding, he drew out the sheets and spread them on the floor of his crude hut. There was not one sheet that was not as blank as when he had first bought it. He snatched up sheet after sheet and hurried outside to peer closely in the brighter light of day, but no amount of staring could bring up the slightest sign of an inscription. After two years of waiting, after three desperate journeys into the close-guarded fortress, after all his hoping and dreaming; it was some time now before Eggborn could fully bring himself to believe that it was not in his fate to succeed in his quest. When realization did come it was followed by grief and despair, and he flung himself to the ground and abandoned himself to weeping.

Hours passed. All at once a sound disturbed him, and he looked up to see, standing a few yards away, the old man dressed in white who once before had guided him.

'It is all in vain,' cried Eggborn to the old man. 'I did as you advised, but it is not in my destiny. The sheets are blank.'

'Of course they are blank,' replied the old man, as if explaining something to a simpleton. 'They are only ordinary sheets of paper. No brush has written on them, no block has printed them, how could they be anything else but blank? I did not say that they would be covered with black ink like a scroll newly written. You examined them in broad daylight—what did you expect to see then? If you really wish to see what the sheets contain, or whether they contain anything, you must wait for a night of full moon. Then you will see what is to be seen.'

Once more the stranger had given Eggborn new heart, and he waited with fresh eagerness for the next full moon only three nights hence. At the end of the third day he climbed to the hill-top above his hut and spread out his precious papers again on the grass. The moon rose full and clear in the sky. On the white sheets, to Eggborn's inexpressible joy, its light brought out the pale grey shapes of characters. He had realized that the moonlight would not last forever and had had the foresight to equip himself with brush and ink. He worked swiftly but with care to fill in with ink the mysterious shapes on the paper. The work was finished long before the moon began to fail. When he had done Eggborn straightened up and cried out to the night sky, 'But I cannot read a word of it! It is in a language I have never seen!'

The reply to his lament came almost at once, out of nowhere, in the voice of the old man: 'Seek out Aunt Piety!'

That was all.

'Who is Aunt Piety, and where must I seek her?' Eggborn demanded; but this time there was no reply. Even now his quest was not at an end, nor would it be until he had found someone to decipher the text he now possessed. Would Aunt Piety be that someone? He could only gather together his belongings and set forth to search for her here, or there, or somewhere in the world of men.

3

The Text

THE priest Eggborn wandered for months by field and village and river and hill in search of Aunt Piety, and although he did not know it he was drifting each day a little nearer to where she was. One morning he woke up with a splendid idea. Leaving the monastery where he had stayed the night he made his way to the near-by town. There he sought out a shop which sold fans, and buying himself a good big palm-leaf fan, he wrote upon it large and clear the words 'In search of Aunt Piety'. This, he thought, would announce his intention to all who saw it. But he did not have to carry it very far. The shopkeeper himself, as he watched Eggborn writing, exclaimed, ' "In search of Aunt Piety", indeed! Tell me, who isn't in search of Aunt Piety?'

'What, do you know her?' asked Eggborn, astonished.

Again the shopkeeper repeated his words. 'Know her? And who doesn't know her? For the past month every wandering monk and Taoist priest in the province has been making his way over here to visit Aunt Piety. It's a fine way to fill yourself full of free rice and vegetables, for she has persuaded Prefect Yang, an old Buddha, to give the biggest feast that has ever been heard of.'

'And where is this Prefect Yang, this Old Buddha of yours?' asked Eggborn.

'Not two days' journey from here,' was the reply, 'in the city at the foot of the Great Hua Mountain.'

For what had happened was that Aunt Piety, as the months had gone by, had grown impatient of the fulfilment of the prediction made to her by the Queen of Heaven. She had repeated it often to herself:

> '*The Aspen will detain,*
> *The Egg will all explain.*
> *Seek not: yourself be sought;*
> *All other search is vain.*'

The Aspen—Yang—had detained her, that was sure. As for the Egg, time alone would show to what that referred. But she began to wonder what was meant by the words, 'Seek not: yourself be sought'. If she was not to seek—and she had no idea what to seek anyway—then clearly she must stay where she was in the residence of the Prefect Yang. But how could one 'be sought'? At last it struck her that whoever was seeking her must surely be a man of religion. In that case, it would perhaps be as well to help him by getting the Prefect Yang to give a great feast and to invite to this such monks and priests as cared to attend; for not even wildfire itself spreads so quickly as the news of a feast among hungry clerics. It had been no difficult matter to convince the Prefect that a lavish feast would ensure a place in Heaven for himself and his wife also. And now it had been in progress for a full month, without any sign for Aunt Piety of anything out of the ordinary, when suddenly one name came to her ears from the servants' report of those who had arrived that day.

The name was Eggborn, and Aunt Piety shivered with excitement when she heard it. For this was he, this was the Egg of the prediction. She herself had been sought, and now was found.

Her eyes flashed; but at once she told herself that she must be careful how she welcomed this stranger. To explain to the Prefect Yang her joy at this priest's arrival would mean revealing the whole story of her meeting with the Queen of Heaven, and of the prediction which contained Yang's own name. Although she herself did not fully understand the grand design in which they were all involved she was certain that its secrecy must be preserved.

Much the same feelings were in Eggborn's breast. He had long come to realize that the old man dressed in white who had helped him over his despair was none other than the White Monkey Spirit himself. Observing the sincerity of his vow, the Guardian himself had helped him penetrate the mystery and had finally brought him to Aunt Piety. But it would never do if, in his relief at the sight of the old lady, Eggborn were to let slip to mere bystanders the knowledge of all that had passed and of the Text of Heaven which lay, even then, wrapped in the bundle on his back.

The quick wit of Aunt Piety saved the situation. After greeting Eggborn with delight she turned to the Prefect Yang.

'A great good fortune for me, which I owe to your generous provision for us all,' she chattered, bowing and smiling. 'This priest is none other than the great benefactor of my previous incarnation, just as you have been in my present lifetime.'

Eggborn could make no sense of these words. But he was quite willing to accept that Aunt Piety should have her own reasons for disguising the strange facts which had brought them together. As for the Prefect Yang, he marvelled once again at the mysteries of the Buddha's Law, and was quite agreeable to Aunt Piety's suggestion that she and Eggborn withdraw to talk over in private their memories from a previous existence.

As soon as they were alone, each gave the other a full account of all that had transpired, and each sighed in wonder over the other's story. Eggborn, rash as ever, was for opening up his precious Text without delay, but Aunt Piety stopped him.

'This is not something that can be done in an hour or a day,' she said. 'We must lay our plans carefully. The study of the Text will take many months. We shall need money to buy the things we shall require. And we shall need help: two are indeed better than one, but three are better than two.' And she told Eggborn of her son, Blackfoot, and of how he still remained in the monastery some miles away where she had left him before her meeting with the Queen of Heaven.

They agreed to send a messenger to summon Blackfoot. The next step was to announce to the Prefect Yang that the two of them, with Blackfoot when he arrived, wished to go into seclusion to meditate and to perfect themselves. This was rather distressing to Yang, for he and his wife believed that they had benefited greatly, and wished to go on doing so, from Aunt Piety's readings of the Holy Books, and stories (often made up on the spur of the moment) of the lives of the Saints. Aunt Piety used all her wiles on him and promised to introduce him and his wife in the Next World; and she backed up her arguments by manifesting herself again as the Bodhisattva Kuan-Yin riding her white elephant on a cloud. And so the Prefect at length not only agreed to her request, he also provided the perfect quarters. These were a suite of rooms about a courtyard some distance from the residence. It was here that the Prefect himself stayed for a week or two each year while he was collecting the rents from his tenants. For this reason the place was complete with furniture, kitchens, food-storage rooms and everything Aunt Piety wanted. It was walled-off and totally secluded and could be entered only by a single gate. This was locked forthwith and the key given to Aunt Piety.

Soon Blackfoot arrived, and Aunt Piety introduced her son to Prefect Yang and his wife and to Eggborn. Blackfoot's first question concerned certain huge jars of wine which stood in a corner, but Aunt Piety decreed that he must fast now and the light died out of his eyes. His second question was after the whereabouts of his sister, and he sighed in wonder when Aunt Piety described how she had vanished from the tomb of the Tang Empress on that night of storm many months ago.

Mother and son, together again, now locked themselves away with Eggborn in the seclusion of their new quarters. In preparation for their studies, Aunt Piety first set about the task of translating the Text of Heaven. 'These sheets of paper of yours,' she told Eggborn, 'are much too big and clumsy. For the translation we shall use little booklets which will be easier to handle.'

Eggborn was therefore despatched to buy small sheets of paper, together with brushes and ink of the highest quality. On his return he took his seat in an attitude of eager reverence, and took down the new version of the Text in good plain Chinese at Aunt Piety's dictation. With each new line he copied, though he might not understand it, he felt an increasing glow of elation; Blackfoot hovered near with frequent cups of tea to refresh them; and in a day and a night the task was completed. Aunt Piety carefully checked the whole work, then took Eggborn's original sheets of mystic writing and set light to each one: for as she said, of such a work it was permissible to have one copy, but certainly not two.

Now all was ready for the commencement of their studies, and at this stage the Prefect Yang paid them a final visit. He was followed by a servant bearing a large box which proved to contain one thousand ounces of silver.

'I hope this will be of help to you during your holy labours,' said the Prefect. 'It should take care of all of your needs for the year.'

In fact the Prefect was not being quite so generous as it may seem; for Aunt Piety had promised that whatever silver he might lend them would be repaid as pure gold as soon as her son Blackfoot had perfected his powers as an alchemist. Yang gave them his final good wishes, then returned to his own residence rejoicing in the thought of the good works which were to take place under his patronage.

Perhaps he would have been more mystified than cheered could he have watched the strange trio as they began their preparations.

First of all Aunt Piety sent Eggborn to bring earth from the four quarters: a handful from where they stood, then a handful from a spot three miles to the east, then from the south, from the west and from the north. Each handful was placed in a bag. To these bags were added objects of every sort and kind: gold and jewels for precious, wood and stone for plain, beans and wheat for eating, coal and charcoal for burning, earthenware for coarse, silk brocade for fine, tea and wine for clear, herbal draughts for cloudy. All these Aunt Piety, after she had bathed and fasted and selected the auspicious day, placed in order between the bags of earth set one foot apart. About the whole a wall was built of new bricks, eighteen inches high. Three bright lamps to burn perpetually were set on top of the wall beneath a yellow awning. In front, on a table, paper likenesses of the gods were set up. This was now the place where the three magicians were each dawn to recite their morning spells and make offerings of tea, wine and fruits.

The next step was to equip with magic powers the paper, ink and brushes which Eggborn had earlier been instructed to purchase. For this were needed seven times seven days of writing charms and reciting spells. At the end of this time Aunt Piety announced that the time had come to summon the generals. These, she explained to the others, were the spirit generals who hover always close at hand but are never seen by ordinary mortals.

'But will they come if we call them?' asked Eggborn; and Blackfoot added, 'What do we do with them when they do turn up?'

'When the ten organs of your body are fully disciplined,' answered Aunt Piety, 'then the ten spirit generals who correspond to these organs have no alternative but to answer the summons. At first they may seem to you fierce and frightening, but you must not show fear; nor, if they appear absurdly ugly, must you laugh. They will do your bidding, but be sure when you call them that your purpose is planned and your desire serious, for they are easily offended and may not come a second time.'

And so they proceeded to write charms and cast

spells, all at Aunt Piety's direction. They went on for seven days, then twice seven. After the thrice seven there came the first signs of response—a faint clang of weapons, a glimpse now and then of a robe. Four times seven, five times seven, and now whole men would come into view, sometimes singly, sometimes in groups, but never stopping. But on the day when the seven sevens were completed all ten of the generals assembled in a room which, although each general was attended by many followers, still somehow did not seem at all crowded. Aunt Piety faced

them, Eggborn and Blackfoot standing to the rear of her.

'We three,' Aunt Piety informed them, 'are the chosen ones of the gods. We are to follow the dictates of the secret Text of Heaven which we hold, and you are to assist us in our tasks. When the work is completed, we will report to your superiors, and you will all be promoted.'

The generals muttered respectful agreement, bowed and withdrew.

But Eggborn, thinking it all over afterwards, was not in the least content with this. He was far too impatient merely to look on as Aunt Piety's assistant, and to wait perhaps for years for the fun of ordering a general about. And so he determined to make a test for himself. He rose from his bed while it was still dark, made his way stealthily to the altar and there performed his incantations. No sooner had he finished than there was a sound like a clap of thunder, and in the courtyard before him stood an apparition which though but dimly seen against the darkness was none the less terrifying. The general's eyes were as big and round as copper coins, his cheeks purple and his whiskers bristling. His armour shone with a gleam of gold, and in his hand was a jet-black banner.

'I am here at your summons, master,' he roared. 'What is your command?'

Poor Eggborn felt his cheeks flush and his heart go bump and had no idea what to say. Frantically he thought back and forth: then his eye lighted on the bare walls of the courtyard.

'There is not enough shade in this courtyard,' he said. He looked round wildly, then pointed. 'Let us have those four tall pear-trees in the Prefect's garden out there moved in here as soon as possible.'

The spirit general grunted assent and disappeared. A few seconds later there came a great rush of wind, flying pebbles clattered on the tiles and one would have thought an army was tramping through the courtyard. When the wind died down and daylight broke Eggborn rubbed his eyes in wonder to see four huge stately pear-trees dwarfing the courtyard in which they now stood.

Aunt Piety of course saw them too; indeed they could hardly be missed. She was furious, and gave Eggborn a thorough scolding.

'I was only trying just this once,' mumbled Eggborn. 'I swear I will not do it again.'

Meanwhile the absence of the trees from Prefect Yang's garden had been duly noticed and reported. Just as the Prefect was scratching his head as to why, when trees had been blown down by the storm in the night, their trunks should not at least be lying there in the garden, old Wang the caretaker came running up to tell him of the four pear-trees in Aunt Piety's courtyard.

'The courtyard is walled in, the gate is locked and I told you never to peep inside,' said Yang hotly. 'How then did you see these trees?'

'The trees are three times as high as the wall, and they weren't there before,' was the reply. 'It is a mystery to me.'

The Prefect realized that this was an effect of Aunt Piety's holy powers and said no more about it, cautioning the servants to keep silence also.

Old Wang soon had another mystery on his hands. The magicians had now begun in earnest the study of the Text. The first secret they learnt was that of the clandestine transport of goods. Thus they no longer needed to ask old Wang for their supplies of rice and vegetables; yet when Wang checked his stocks he discovered that they were steadily diminishing. This again he reported to the Prefect, who again told him not to reveal the fact outside.

At last the three years which Aunt Piety had predicted for their task were nearing an end. And each of the three confederates had perfected his seventy-two magic skills. Each could ride the clouds, hide under the earth, move mountains or turn rivers. Each could take on any form he chose, or make beans into armed men, or stay unharmed by fire, flood, sword or spear. These are after all but mean tricks, and not such wonders as a Bodhisattva might perform. Yet they are tricks that could prove dangerous in the wrong hands; and this was why the Jade Emperor had set the White Monkey to watch over the Text when once it had been inscribed within mortal reach. The Monkey Guardian himself had added certain instructions at the close of the Text. These obliged whoever came into possession of it to repeat each year a solemn vow never to use its secrets to cause harm to mankind, on pain of awful penalty. Poor Eggborn, however, had been in much too much of a hurry when he copied down the Text to trouble about the tiny characters appended. In consequence of this, he and his companions, their study ended, now felt themselves perfectly free to practise their arts. And in fact, in the course of time, great trouble was brought to many men, warfare raged throughout the north of China and the names of Aunt Piety, Blackfoot and Eggborn himself were sounds to be dreaded and detested.

For the moment, however, the three of them discussed pleasantly how before leaving they should repay the Prefect Yang for his many kindnesses.

'I'm going to give him a tiger to guard his estate,' said Blackfoot.

'We received such a welcome gift of silver,' said Aunt Piety, 'I think I should repay this by turning this nice big rock here into gold.'

'Good!' cried Blackfoot. 'My tiger can guard your gold rock!'

'Perhaps he might welcome a memento of the three of us,' suggested Eggborn. 'I'll have a first-rate sculptor carve our likeness in stone.'

All agreed that these would make three handsome gifts. Aunt Piety muttered a spell, then spat on the rock. Her spittle turned into a mist which enveloped the rock, and when she rubbed it with her hand the whole surface gleamed with the gleam of solid gold. Blackfoot cut the figure of a tiger out of paper, muttered a spell, tossed it into the air and cried 'Quick!'—at which the paper cut-out fluttered to the ground and at once bounded up again, now a flesh-and-blood tiger striped and roaring.

> '*Tiger, tiger, I thee constrain,*
> *Guard the gold rock with might and main.*'

—so Blackfoot commanded the tiger, which with a wave of his long sleeve he then turned back to its paper form. At the bottom of the gold rock was a cavity, and he tucked the paper shape into this.

Now it was Eggborn's turn. He captured the soul of a cunning sculptor whom he set to work on a group of the three of them. In one night the work was completed, and the likeness was such that Eggborn felt uneasy and no longer sure of his own identity to see himself sitting there.

Aunt Piety was reflecting on the fulfilment of the prophecies concerning the Aspen and the Egg. 'It was the Queen of Heaven's will that brought us here together,' she said. 'And I think now of her further instruction: that she would have need of me twenty-eight years hence in Peichou. What we must do now is disperse, each to wait in his own way for the summons which must come. But when it does come, let neither of you attempt to disobey it, for it is the decree of the Queen of Heaven.'

With these final words Aunt Piety ascended into the blue of the sky, from whence she beckoned to the two who remained below. Eggborn then tossed his staff into the air, where it changed into an endless bridge of gold, and off he went, striding up the bridge. But Blackfoot declared, 'I'm for a spell in the Bottle Paradise,' and finding an empty wine-jar in a corner he set it firmly on the ground, called out, 'Here I come!' and disappeared into the mouth of the jar.

And so they were gone. Neither the Prefect Yang nor any of his servants had ever ventured to disturb their privacy, and now it was some time before the fact of their departure was discovered. But at length an inquisitive gardener set a ladder to the wall, peeped over and was shocked to see the three of them sitting there together in the stillness that only death can bring. Deeply disturbed by his report, the Prefect Yang unlocked the gate and entered their courtyard. His relief can be imagined when he found his friends not dead, but magically modelled there in stone. He recognized that, their holy labours completed, they had deemed it unfitting to take formal leave; and he was touched as well as awed to find the great rock of gold they had left as a parting gift. Men were summoned to move the rock into Yang's own residence: but no sooner was the first hand laid on the rock than a ferocious tiger leapt out of nowhere and sent the crowd flying with its roars. When at length a few hardy souls plucked up the courage to look again, no sign of a tiger was to be seen. But a second attempt to move the rock met with the same response, and from then on the rock was left where it stood, by the side of the lifelike sculptured group, as a memorial to the strange trio who had lived and worked there.

And indeed, not until twenty-eight years and more had passed did any change take place. And then one day a casual visitor to the now-famous site was to find a piece of sculpture all crumbled away, a rock without the slightest trace of gilt about it, and at its foot the crumpled shape of a tiger crudely cut from a piece of paper.

4

Eterna

SINCE the night, years before, of her dramatic meeting with the Queen of Heaven, Aunt Piety's thoughts had turned many times to her missing daughter. Where was she, and how was she preparing herself for her future destiny? No message had come from her, no word had reached Aunt Piety's ears from any source at all which mentioned her name.

Now this was not surprising, for the girl was no longer even among the living—in her own person that is to say. She had died and had been reincarnated, and was now a young girl of thirteen. The change had come about through her own impatience. On the night of the storm it had been the rushing black wind which tore her from her mother's side and whirled her away through the air. As she spun, dazed with terror, in the wind's grasp, a voice rang clear in her ears, twice repeating the following words:

'Have no fear over what has been;
In time to come you shall be Queen.'

When at last the black wind set her gently down on earth again these words were all that filled her mind. She found herself in the Eastern Capital, in a narrow street at the end of which she could see the gleaming tiled roofs of the royal palace. By stealth and cunning and the exercise of her skilful tongue she made her way into the palace and into the apartments of the royal Prince himself. But her stay here was to be short, nor was she to leave alive. Her plan was to make the prediction she had heard come true all the more speedily, by winning her way into the affections of the Prince so that in the future, when he came to the throne, she might hope to be his Queen. But the guardian gods of the royal palace were not deceived by her fair human form. They knew her very well for what she was—a vixen of magic powers. And a vixen could bring nothing but harm to a mortal Prince. No sooner had she entered the princely chamber, at dead of night, than a lightning flash blasted the room. When the guards came running to see what was amiss they found the Prince sleeping peacefully, unharmed; but in a corner by the wall lay the small, dead, scorched body of a she-fox.

The carcass was taken out and burned. But the soul of Aunt Piety's daughter had already left to go winging its way in search of rebirth. It found its place in the family of a rich shopkeeper, a man named Hu, himself also a resident of the Eastern Capital. To this man's wife, not long after the strange events in the palace, was born a baby girl, and to this girl was given the name Eterna.

And now years had passed and Aunt Piety, having taken leave of Blackfoot her son and of Eggborn the priest, set herself to discover the welfare of her long-lost daughter. By her magic arts she traced the girl's soul to the body of Eterna, in the Hu household in the Eastern Capital.

'She seems a pretty enough child,' said Aunt Piety to herself, 'and let us be thankful for that. But she will need some training in the practice of sorcery if she is to play her part in the scheme of things at Peichou.'

But the more Aunt Piety reflected, the more doubts she had of attracting the girl's interest towards the magic arts. For the great appeal of magic is to those in need, to the starving or the desperate or those beset by perils. The family of the merchant Hu was rich and respected and in need of nothing: what use had Eterna for Aunt Piety's wicked craft?

Well, this could soon be remedied. The old woman's first move was to bring down fire on the merchant's shop and his warehouse and his residence itself. After that it needed only a few lesser blows to reduce him and his wife and daughter to the condition of paupers. It was mid-winter, snow lay on the ground, and before many days were out the whole family faced starvation. Hu himself put on such of his thick clothes as he had not already pawned to buy rice, and went out to beg for help from his former wealthy friends. His wife and the girl Eterna sat on in the miserable cottage they had been brought to; but at length they could bear the pangs of hunger no longer.

'My child, you must go out and buy a few wheaten cakes for us to eat, or we shall not last till your father returns,' said Mrs Hu sadly. 'There are a few coppers under my pillow—they are all we have left. Take them and buy what you can with them.'

The most careful search revealed only eight coppers. Clutching these in her hand Eterna went out, in her thin dress and tattered slippers, into the snow. In the city streets the snow had turned to a muddy slush, and turning a corner Eterna slipped and fell full length. When she picked herself up her dress was wet through and filthy, and the eight copper coins were scattered all over the ground. She found seven, but after a long and painful search had to give up the last one for lost. Wet and shivering, she went on to the cake-seller's stall, where she asked for seven cakes and handed over her money.

'This one's all bent, I can't take it,' said the cake-vender; and he gave her one of the coins back, and only six cakes with it.

Threading her girdle carefully through the hole in the centre of the rejected coin, Eterna turned back with her cakes. But she had not gone far when she came face to face with an old woman, a hag bent and ancient who leaned on a stick and had a basket on her other arm.

'What have you been buying there, my dear?' asked the hag with a toothless grin; and Eterna showed her the cakes.

'I had neither supper last night nor breakfast this morning,' went on the hag. 'Won't you give one to me?'

Eterna well recalled that she herself had had neither supper nor breakfast, but nevertheless she handed over one of the precious cakes. The old woman beamed her approval of such kindness; but in fact it was not kindness but sheer terror that had moved Eterna's rather selfish little heart.

Then the old woman went on, 'You know, my dear,

one cake isn't nearly enough to still my hunger. Why not give me all of them?'

But Eterna told her how her mother and herself were also starving, and recited the whole sad story of how she had lost one coin and had another rejected as bent and worthless.

'Then you may keep your cakes,' said the old woman. 'I didn't really want them—I was only testing you out, to see whether you had a kind heart. I find that you have, for you gave me one of your cakes; so now I'm going to do something for you. Let me have that bent coin for a second.'

Eterna untied the coin from her girdle and handed it to the old woman, who took it in her hand and blew on it and handed it back. Eterna looked—it was bent no longer, but flat and shining and as good as new.

'Why, that's marvellous,' Eterna exclaimed. 'Can you teach me to do that?'

'Oho,' said the hag, 'that's only a feeble little trick. If you really want me to teach you something, just take this book and read it'—and she took from the basket on her arm a little book all wrapped in crimson cloth. 'There are lots of things in this book,' she continued, 'and when you are in great need you may open it and learn some of them. If there is anything you find hard to understand, just call out "Aunt Piety", and I'll come along and help you. But whatever you do, don't let anyone know of this.'

Eterna took the book and thanked her, but when she had walked away a few steps and looked back the old woman was nowhere to be seen. Eterna felt there was something special about the coin that had been bent but now was as new. On returning home she therefore told her mother that she had lost two of the coins in the snow. They ate the cakes between them, and then when Mr Hu came back from his expedition were delighted to learn that he had been given three hundred coppers by a man who not long before had been a servant in their employ.

But even three hundred coppers lasted only a day or two, and then things were as bad as ever. 'This is surely a time of "great need",' Eterna told herself. 'Let's see what this little book can really do for us.'

Mindful of the old woman's instruction of secrecy, she waited until dead of night, when she rose from her bed, went into the kitchen and opened the book in the moonlight at the window. There she read what was written on the first page:

'*Multiplication of Coins.*

'Take one string and thread on to it one copper coin. Place on the floor and cover with a vessel. Holding in the hand a bowl of water, recite the formula seven times. Spurt one mouthful of water over the vessel and issue the command "Quick!" On raising the vessel one whole string of one thousand coins will be found beneath.'

Eterna followed the instructions exactly, using the coin which the old woman had returned to her and the ribbon with which the book had been tied up, and reciting the formula which was given at the end of the page. She lifted the jar she had used for a cover—and lo and behold, there like a greeny-brown snake lay a thousand copper coins threaded on the ribbon!

Eterna was overjoyed: but what to tell her parents when they asked where the money had come from? She cautiously opened the door and laid the money on the snowy ground just outside, so that it might be thought some rich benefactor had left it there for them during the night. And this was exactly what Mrs Hu did think when she opened the door and found the money the following morning. Mr Hu was not so easily satisfied and worried his poor head over the trouble that might ensue from such a mysterious gift; but his wife told him not to be so silly, and when he had spent the money on firewood and rice and vegetables he felt a great deal happier. The next night Eterna took out a string and a coin which she had carefully made ready during the day and repeated the experiment with just the same success as the first time. Again Mr Hu was puzzled, but again he was quite willing to be pacified by his delighted wife.

Soon came a day when, the gift money used up, Mr Hu had gone out again to seek the help of old friends, whilst his wife was round passing the time of day at the neighbour's. Alone in the house, Eterna determined to open her book again and see what exciting new magic might be on the second page. And indeed it was a new trick: '*Multiplication of Rice*'. 'Praise be to Heaven and Earth,' thought Eterna, 'as long as we can "multiply rice" we need never go short of something to eat.' The instructions were much the same as before. Eterna set the rice-cask, in which but a few grains remained, on the floor and covered it up with her jacket. She recited the new formula, spurted the water and cried 'Quick!' But alas, the rice-cask was old and feeble and filled up so quickly with the magic grain that it burst and spilled a cascade of rice all over the kitchen floor. A small hillock of rice now filled the centre of the room and before Eterna's startled eyes continued to grow. She screamed in dismay, so loudly that her mother heard her from the house next door and came dashing home to see what was the matter.

With the intrusion of an outsider the magic ceased;

but the girl was still left with a kitchenful of rice to explain away. In desperation she mumbled a tale of a big strange fellow who had marched in and swung down a huge sack of rice from his shoulder, emptying it into their cask, which had burst. She had been so startled, she said, that she screamed, whereupon he had left without a word.

Mrs Hu was quite ready to swallow this, but with her husband when he returned that night it was a different matter. Two strings of coppers were mystery enough, but a big fellow loaded with rice—it was altogether too much! He demanded the truth from his daughter, and threatened to beat it out of her with a stick if necessary.

It was in fact the beginning of a long struggle between Mr Hu and his strange daughter Eterna. Faced with the prospect of a beating she told her father all about her meeting with the old woman and about her experiments with the aid of the book she had been given. Mr Hu was horrified. He well knew the attitude of the authorities towards witches and sorcerers of any description, and he knew the risks one ran by being in any way concerned with 'dubious practices'. Without further ado he thrust the book into the stove and watched it burn, sent Eterna to her room and ordered her to forget everything that had happened.

Which was all very well: but it provided little comfort for hungry stomachs. Before many days were out the gift money had been spent and the magic rice eaten up and they were in a worse plight than ever. Now Mrs Hu began to scold her husband for treating the girl so harshly when after all she had only been trying to help; and to Mr Hu himself the anger of the authorities, if ever they were to find out, began to seem less terrible than the pains of a dry mouth and an empty belly. And so it came about in the end that he went to his daughter's room, humbly apologized for scolding her and for burning the magic book, and begged her to try to remember the methods she had used for multiplying coins and rice.

Eterna, meanwhile, had been far from content merely to lie on her bed and mope. She had called as instructed to Aunt Piety for help. The old sorceress taught her three things immediately: how to leave the house without opening door or window; how to transform a stick into the sleeping body of herself, to deceive her parents when they looked into her room at night; and how to ride a bamboo staff through the air to join Aunt Piety herself in a place above the clouds. There, night after night, the old woman had been instructing Eterna in the further exercise of magic.

Calling now on Eterna in her room in the day-time, Mr Hu found her sleepy-eyed and unable, so she said, to remember her former tricks. 'But I will try, Father, if you wish,' she said. 'Just sit down on that stool there while I say the words.'

Mr Hu sat down. Eterna spoke a spell and cried, 'Quick!'—and the stool with Mr Hu on it flew up in the air. But for the ceiling it would have soared high in the sky; as it was, poor Mr Hu cracked his head on a beam and screamed to be brought down. 'I'm sorry,' said Eterna, 'this seems to be the only trick I remember, and it isn't much good for filling a hungry belly!'

But at last she repented, let her father down and told him to prepare two strings with one coin on each for the purpose of multiplication. Mr Hu went happily off. But as he prepared the strings he asked himself why should he stop at two? Why not get a hundred strings and have something to spare? He did not have a hundred lengths of string in the house, and so he went round to the oddments shop on the main street.

'Certainly I have a hundred lengths of string,' said the shopkeeper. 'But what do you want them for?'

'For threading coins,' said Mr Hu.

'Congratulations,' said the shopkeeper. 'You must have come into a great deal of money. Here are the strings—that will be twenty coppers, please.'

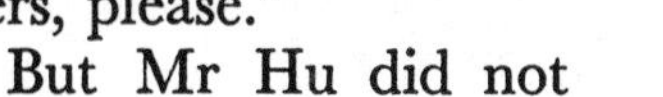

But Mr Hu did not possess twenty coppers. He promised to bring the money to pay for the strings in half an hour's time; he tried to pawn his jacket in exchange for them; but the shopkeeper was adamant, for he suspected some mischief from this fellow who wanted a hundred strings for threading coins but did not have twenty coppers to pay for them. And in the end Mr Hu had to return crestfallen to his house and make do with a mere two strings.

But two strings today, and two tomorrow, and a kitchenful of rice whenever it was wanted! From that time forward the Hu household went short of nothing, and as Eterna's skills progressed and her father's greed increased he grew wealthier than ever he had been before the calamities struck. He moved his family from the miserable cottage into a fine mansion with many handsome garden-courts. Nor did he allow himself to worry any further over the possible consequences of his daughter's mysterious talents: until one fateful day when he was looking for her in connexion with a little matter of 'multiplying' some fine brocade for sale in his shop.

He could not find Eterna anywhere in the house, and began to look in one courtyard of the mansion after another. At last he reached one of the smallest and most secluded, and there he saw Eterna. She was kneeling on the ground with a bowl of water before her. In her hand she had a red gourd-bottle, and as Mr Hu watched she took out the stopper and emptied the gourd. On to the ground rolled a large number of tiny red beans, followed by a cascade of short pieces of chopped-up straw. Eterna spoke some words, spurted water from the bowl over the beans and straw, and cried 'Quick!'—and at once the courtyard filled with warriors, men three feet high, all red; red helmets, red armour, red skirts, red lances, red banners; two hundred of them, each one seated on a red horse and all ranged in battle formation. Now Eterna, before Mr Hu's astonished eyes, took up a white gourd, out of which she poured white beans and another heap of chopped-up straw. Again she muttered and spurted water and cried 'Quick!', and this time there appeared an army, another two hundred, of warriors all in white, mounted on white steeds and drawn up in line of battle. Eterna took a pin from her hair and changed it into a jewelled sword; then, waving this, she gave the order to attack and sat back to watch the two contending armies surge back and forth. It was only a tiny courtyard, yet it did not seem crowded even though fully four hundred troops were filling it with the din of warfare.

Mr Hu, still unseen by Eterna, could stand it no longer. Surely a terrible end must come from such deeds! On an impulse he ran back into the house, seized a huge, sharp chopper from the kitchen and returned to the scene of battle. By now Eterna had grown tired of the game and had returned the warriors to beans and the beans to the gourd-bottles. Mr Hu ran up behind her and without warning swung the chopper through the air. It cut clean through Eterna's slender neck and her head dropped to the ground. In anguish, Mr Hu dragged body and head beneath some bushes and left the courtyard, the gates of which he locked behind him. All that day he was heavy-hearted, and when night came he called his wife to hear his tragic story. His wife was astounded. 'But when did you do this dreadful deed?' she asked.

'Today,' he replied, 'in the middle of the afternoon.'

Mrs Hu smiled with relief. 'Then there is someone I should like to show you,' she said. And she led him into Eterna's room, where their daughter lay peacefully sleeping.

Eterna opened her eyes. 'You hurt my neck, Father,' she said. 'You should not have struck so hard.'

Mr Hu realized that his weird daughter was not to be got rid of so easily. Yet if she continued to live in his house there could not fail to come at last some terrible disgrace which would involve them all in ruin. There was only one way out. The girl must be married off, then at least it would not be himself and his wife who were caught up in the harm that must befall.

A husband then was sought for Eterna, and the instructions given to the go-betweens made them open their eyes in wonder. Mr Hu had hit on the clever scheme of marrying her to a half-wit, a man so simple that he might never realize the strange truth about his bride. Eventually such a man was found, an imbecile named Chiao, and Eterna passively permitted herself to be married to him. He had not the intelligence to be surprised by anything she did, and thus she was free to practise her tricks quite openly in her own home. All went well until one summer when the weather turned hot and sultry day after day. Eterna suggested one evening that they go up to some high place to escape the heat, and her husband agreed: indeed, the poor fellow never did anything but agree with anything anyone said. They seated themselves together on one stool, Eterna spoke some magic words, and in a twinkling the stool rose in the air, crossed the city and deposited them safely on top of a watch-tower, high above the city wall. There they sat and enjoyed the cool breeze, Eterna fanning herself with a large silk fan.

Below them, two watchmen were making their rounds of inspection. 'Hello,' said one to the other, 'what's that up on the tower?'

'Crows,' said the second.

'I don't think it is.'

'Then why is one flapping its wings?' returned the second.

'Wings, rubbish!' said the first. 'That's a woman waving a fan.'

'Well, woman or crow, it's no right to be up on our tower. Let's see how it likes the taste of an arrow.'

The watchman raised his bow, took aim and shot, and down fell poor Chiao the imbecile with an arrow through his thigh. Here was trouble indeed! Eterna at once instructed the stool to return her to her home, but when the wounded Chiao was summoned to the magistrate's court and questioned some inkling of the truth came out, and runners were dispatched to bring in the half-wit's wife.

In desperation Eterna called on Aunt Piety for help. And now she learned to her joy that the time had come for her to leave the Eastern Capital and join Aunt Piety in her dwelling. She set out alone on the road to Chengchou. Following Aunt Piety's command she came to a certain place where stood a well. Eterna climbed on the rim of the well, and jumped.

She was not surprised to find no water in the well. She did not fall, but floated gently down to land safe at the bottom. It was dark; but waiting maidens guided her out of the darkness into a shining path, cobbled with precious stones and lined with rare trees. The path led to a palace of such magnificence as has never been seen on earth; and here, in a lofty hall, Aunt Piety robed in splendour awaited the return of her long-lost daughter.

5

Recruits

IN the city of Chengchou, not far from the well down which Eterna jumped to join Aunt Piety in her palace, in a crowded alley lived a pieman whose name was Jen. He had two good friends, Chang the butcher and Wu the noodle-seller, and each of the three was big and strong and expert in his trade. After a day's baking Jen the pieman set out early the next morning for the cross-roads where he usually found the most customers. He greeted his fellow-venders who were also just arriving, spread out his pies and rolls and sat down to await the day's passers-by. He had not been there many minutes when a sound reached his ears, the sound of a rattle such as is carried by Taoist beggar-priests. On looking up he saw approaching such a priest, a truly comic sight: short, dirty, a tattered robe streaked with grime, and a rag round his head above which a greasy mat of uncut, unkempt hair rose stiffly.

The stranger came nearer, limping heavily, one foot high, one foot low.

> *'I bring you luck,*
> *I bring you wealth,*
> *Give me a copper*
> *To keep me in health.'*

This was his chant as he stopped before Jen's tempting display. 'Careful, reverend sir, or you'll trip over a rat's tail,' Jen scoffed. 'You're a bit early in the day, aren't you? I haven't sold so much as a steamed roll yet, I've no left-overs, not even a copper to give you.'

'Then how do you sell your rice-balls?' asked the beggar, who was, of course, Blackfoot.

'Two coppers the big ones, one copper the small,' answered Jen. 'Here you are, give me two coppers and you can have one of each.'

'The big one's for my mother,' said Blackfoot, taking them in his hand. Jen noticed that his fingers were black with filth. He kneaded the white rice-balls until they turned grey, then said, 'My mother is eighty—how do you expect her to eat things as hard as this? Make it a dumpling instead.'

'You've made sure I can't sell them to anyone else,' Jen grumbled. Nevertheless he put them aside in the tray for left-overs, took a dumpling from the steam-basket and handed it to the priest.

'What's inside?' asked the priest, clutching the dumpling tight in his grubby fist.

'Best quality beef,' replied Jen.

'My mother doesn't eat meat. Give me a fried cake instead.'

To be rid of him Jen gave him the fried cake, putting the ruined dumpling again with the left-overs. But it still wouldn't do, he must change it for a steamed roll. 'Be off with you!' shouted Jen, exasperated. 'All I've had from you is two coppers and you've spoiled three of my wares. Keep the fried cake you've got and leave me in peace.'

Blackfoot's only response was to blow a puff of breath over Jen's stall. Then he walked away, and the pieman sighed with relief. But when he looked again at his stall his eyes nearly popped out of his head: for every one of his dumplings and rolls and cakes was as black as a cinder. 'It's that priest!' cried Jen. 'Just wait till I catch him!' And he dashed out in pursuit, hurriedly calling to his fellow-venders to keep an eye on his stall.

He ran in the direction the priest had taken but he ran for a long way without catching up with him. Still furious, he was on the point of giving up and going back when his ears caught again the sound of the priest's rattle. He made in that direction, and found a sizeable crowd of people gathered round a butcher's stall.

'This is Chang the butcher's place,' said Jen to himself. 'But what are all these people doing?'

He pushed through the knot of people and found in the centre an old lady lying on the ground while a boy, evidently her son, tried to revive her. In the end she rose to her feet and stumbled away, leaning on the boy's shoulder and moaning. The crowd dispersed, and Jen greeted Chang the butcher and asked, 'What was all that about?'

'What a business!' replied Chang. 'Just now a lame Taoist priest came up chanting something about

"*I bring you luck,*
I bring you wealth,
Give me a copper
To keep me in health."

—I told him he was too early and asked him whether there weren't any windows in his house. He got annoyed: "If you've nothing for me, well and good, but don't laugh at me," says he. Then he came right up against my stall. I had a pig's head on the front that was ordered by old Mrs Ti, and this priest looked at it and stroked it and muttered something—then I chased him off. Well, along came Mrs Ti to pick up her pig's head, but as soon as she touched it it opened its eyes and bit her! She fell down in a dead faint and I thought I had a corpse on my hands, but fortunately she seems all right now.'

'That priest is a monster,' said Jen the pieman, and he told Chang what had transpired at his own stall. Chang roared with rage, and was more than ready to join Jen in his pursuit.

They chased hither and thither for an hour or more without sign of the priest, until they came to the shop of Wu the noodle-vender. Here was fresh commotion: Wu was just on the point of thrashing his assistant, a young lad whom he held in a grip of steel.

'What's the lad done?' asked Jen and Chang.

'Why, you'd think by this time he could have got a fire going,' answered Wu. 'Here's all my customers come and gone and I haven't boiled a single panful of noodles yet because this lazy lout can't get the stove going!'

The same thought struck both the pieman and the butcher at once.

'Have you had a visit from a Taoist priest,' they asked Wu, 'a filthy, lame creature chanting something about "Give me a copper to keep me in health"?'

'Why, yes, now you come to mention it, such a fellow did come along earlier. I asked him if he was afraid the demons would get him if he came later in the day, told him I'd nothing for him and sent him packing. He went without a word, but he blew a puff of breath at my stove—oh, so that's what happened! And here was I going to knock the breath out of this poor lad here!'

No sooner had Wu finished his tale than the three of them heard close at hand the sound of the beggar-priest's rattle. As one man they dashed out. The lame priest was just in front of them and they ran to catch him.

But as they ran, so did the priest, lame or not, faster and faster so that they came no nearer to him. They slowed down: so did he; they ran faster again: so did he. 'He must get to where he's going some time,' said Chang. 'Then we'll catch him.' Without realizing it they followed him seven or eight miles beyond the city wall.

At last they came to a Buddhist temple and the priest dived for shelter inside. 'Now we've got him!' the three told each other, and they followed into the temple. They were just in time to see him disappear into the main Hall of the Buddha. Into the Hall they dashed. At the far end was a giant image of the Buddha. As they watched, the priest jumped on to the Buddha's hand, climbed up to his shoulder and threw his arms about the great head. 'Lord Buddha save me!' he shouted. He tugged at the head—which came off and crashed down to the floor. Then, up and over, and Blackfoot had scrambled down into the belly of the Buddha.

'Now we've got you!' cried Chang the butcher triumphantly, and he jumped up on the Buddha's hand, clambered up to the shoulder and stood on the edge of the gaping neck staring down into the blackness. But as he stood, a hand came up from the belly of the Buddha and seized Chang the butcher's leg. He toppled over and disappeared inside the statue.

Taken aback but still undaunted, Jen the pieman declared that the only thing for it was to take a look and see what was happening. He in turn clambered up on the Buddha's shoulder, looked down into the black inside and called out to Chang. No reply came, but the hand reached up again. 'Oh, reverend sir, forgive me and I'll chase you no more,' cried Jen in desperation. 'I was coming to ask you how many dumplings and rolls you wanted, and I'll send them to you.'

But it was to no avail: the hand closed round the pieman's leg and down he went. Now it was the turn of Wu the noodle-vender. How could he let his two comrades down? He climbed on to the hand, clambered up on the shoulder and stood there, his knees quivering like bean-curd. 'I'll get down again,' he said to himself, 'and find a big stick and smash the whole image to let them out, then we shall see.' But before he could move he felt an arm come round his waist, he was lifted off his feet and down he went in his turn.

It was pitch-black inside the Buddha's belly. Suddenly his foot hit against something and there came a cry, 'You're treading on me!'

'Who's that!' asked Wu.

'It's Jen the pieman,' was the answer.

'Where's Chang?'

Chang's voice replied. But there was no sign of the priest. To their surprise the three men found that they could move about quite easily, there was plenty of room. They ranged themselves shoulder to shoulder and managed to walk forward a few steps, though still in inky darkness. They went on walking, a good half-mile, and gradually it began to grow lighter. They came to a huge stone gate. As they approached it swung open, revealing on its other side a fair landscape with hills and rivers and woods and fields. Through this they walked on for several miles until they came to a little farm, where an old woman came out to greet them.

She was most courteous, and as by this time they were hot and parched and hungry they tried to buy food and drink from her. She would not hear of payment, but set before them a variety of dishes and a large jug of wine. They soon felt much better. Before long they made ready to continue their search: when in at the gate, unconcerned, walked the Taoist priest they had all this time been pursuing.

The three jumped up to set about him. But the old woman restrained them: it was her son! Indeed it was Blackfoot, and the woman was his mother, Aunt Piety herself. They told her the story of the mischief he had caused, and she scolded him and made him apologize. The three men could hardly do other than accept his apology in view of the kindness they had received from his mother, but still they seemed extremely dissatisfied.

'To make up to you for all your trouble,' said Aunt Piety, 'I will ask my son to show you a few of the magic tricks he has learnt. You are all men destined for greatness, and a trick or two may come in useful in the future.'

Jen, Chang and Wu were most excited to hear these words. They watched attentively as Blackfoot untied from his belt a gourd. He held it in his hand, muttered some words and cried 'Quick!' Instantly from the mouth of the gourd there poured a great flood of water, the ground about them became a sea of surging billows. 'Bravo!' cried the three companions.

'Now I'll put it back for you,' said Blackfoot, and wave by wave he returned the water to the gourd. Now he issued a fresh command and cried 'Quick!'—and tongues of flame leapt out of the gourd so that the whole world about them burned with fire. Again the three shouted 'Bravo!', and again Blackfoot led the flames gently back into the gourd.

'Brother priest, would you give the gourd to me?' asked Chang the butcher. And at Aunt Piety's wish Blackfoot gave him the magic gourd.

'Here's another trick,' said Blackfoot. He took a piece of paper and cut out with scissors the shape of a horse. This he laid on the ground, then spoke a spell over it. He cried 'Quick!' and the paper shape stood up, shook itself and turned into a magnificent white steed. Blackfoot leapt on its back and rode off, when soon the horse soared up into the blue. It flew in a great circle round the sky before returning to land safely at their feet. Blackfoot dismounted, and the horse immediately fell down, a mere paper cut-out once more.

'Which of you gentlemen would like the horse?' asked Blackfoot.

'I would,' said Wu the noodle-vender eagerly, and Blackfoot gave it to him.

'Now what about this third friend?' asked Aunt Piety. 'What are you going to show *him*?'

'I can't think of anything just at the moment,' said Blackfoot. But as he spoke, a new person walked out of the house towards them. Aunt Piety introduced her daughter, Eterna, to the three friends.

'*I* have something for this gentleman,' said Eterna; and taking a stool she changed it, with a spell and a shout of 'Quick!' into a huge white-browed tiger which carried her high up into the sky. Then, 'Down!' she cried, and the tiger returned her safely to earth and changed back into a stool once more. 'Would you like to learn this spell?' she asked Jen.

'Indeed I should,' he answered, and to his delight she taught him the magic words.

'Now all of you practise your new tricks,' said Aunt Piety. When they had all three successfully done so, she went on, 'I have but one request to make of you. In time to come there will be need of you in Peichou. You must come when I summon you, then all of you will share in the greatness which will be ours.'

The three friends gave their assent. But before Blackfoot escorted them back to their homes there was one last surprise in store for them. A Buddhist monk in a saffron robe and with gold rings through his ears approached, greeted them and asked what mischief they had been engaged in with the flames and flood and the flying horse and tiger he had seen from a distance.

'Do not be angry with them, Father,' said Aunt Piety. 'These are three new friends, and my children have been teaching them some little tricks.'

'Then let me see their little tricks,' said Eggborn.

The three men obliged, and the gourd and the paper

horse and the stool performed their magic once again. But Eggborn was not impressed. 'These are very minor marvels,' he said.

Chang the butcher was stung by this. 'Then let us see you do better, reverend Father,' he said.

Eggborn said not a word. He merely stretched out his right hand, the fingers open wide. From the tip of each finger a ray of golden light shone forth clear up to the sky; and high at the end of each golden ray appeared the figure of a Buddha in splendid majesty.

Jen, Chang and Wu were speechless, and bowed in obeisance before the greatest wizard of them all.

6

Revolt

THE sorcerers were assembled, for the time was at hand. The events in Peichou of which the Queen of Heaven had spoken long ago were about to be set in motion. Aunt Piety had possessed herself all these years in patience. Now she acted, to set in train deeds which were to lead a whole province to disaster.

She sent Eterna to Peichou. Dressed in the white of mourning, a basket over her arm, Eterna was pretty enough to make many a man turn round to stare at her on the street. The bolder ones asked her where she was going. 'I have some things to sell,' she replied, 'just to earn a few coppers to support me now that my poor husband has passed away.'

'What is it you have to sell?' they asked her.

'Just clear a space for me in this busy market-place and I'll show you,' she said. In no time a space was cleared and Eterna set down her basket. She took out a bowl and asked a bystander to fill it with water. Curious to see what she was up to, he returned a moment later with the water. Eterna now took a knife and scraped some loose soil from the ground; on this she poured water, then from the mud fashioned the shapes of ten candles. 'Now,' she said, 'who will give me three coppers each for my candles?'

The little crowd about her laughed aloud. 'Mud candles!' they cried. 'What use are they?'

But Eterna begged for the use of a torch from a near-by stall and touched it to the candles, which took light and blazed with a brilliant light. Astonished, the bystanders fought to buy these marvellous candles for three coppers apiece; but when Eterna had sold her little stock of ten candles she picked up her basket and left.

The next day she was there again, and the crowd was bigger; and bigger again the day after that, and so on until the news of her spread all through the city. It reached at last the ears of the man she was seeking. This man was a captain of the provincial army, and his name was Wang Tse. He was over six feet tall, broad and strong and skilled in the use of all weapons. His father and mother had died when he was young, he had led a wild youth and it was predicted that he would come to no good. Now this day, idling on the streets, he decided to see just how pretty she was, this young widow who sold marvellous candles for only three coppers. He forced his way through the crowd about her. More than all the rest he was taken by her bewitching face. He bought up all ten of her day's batch of candles and gave them away to the crowd; then, when Eterna took up her basket and left, he followed her at a distance.

She led him through the city, out of the gate and a mile or two through the fields beyond. So far did she go that he was on the point of giving up and going back: but when he looked round, the road behind was not the road he had come. On all sides steep mountains blocked the view; he was hopelessly lost.

Suddenly Eterna was at his side, laughing. 'What a time I had to get you to come!' she said. 'Now you must be introduced to my mother.' All at once he was in a mansion, and there before him stood a dignified and finely dressed old lady.

'Greetings, Captain Wang,' said Aunt Piety, and Wang Tse bowed. 'I had you brought here because you are alone in the world and in need of help. Such a fine man as yourself needs a worthy wife, and I propose to marry my daughter to you.'

Nothing could better have suited Wang Tse's wishes, and though in a daze he gladly allowed himself to be led, with the assistance of the bride's brother, Blackfoot, through an elaborate wedding ceremony. It was such a wedding, with sumptuous robes and rare meats and whole bevies of lovely ladies-in-waiting, as would have befitted royalty; but when Wang Tse commented on this, Aunt Piety replied, 'That is only right, for royal you shall become, the ruler of all this land.'

'But I am only a humble captain of the provincial army,' protested Wang Tse. 'How should I build myself a throne?'

'With our help you may do anything. My daughter, your wife, has a hundred thousand troops awaiting your command.'

Wang Tse laughed, 'A hundred thousand troops would need great granaries to feed them and arsenals to arm them. I don't see any sign of such about here.'

'My troops need neither food nor other supplies,' said Eterna. 'Let me present some of them.' And she took two little boxes, and from the beans in one and the chopped-up straw in the other brought a squadron of cavalry two hundred strong and fearsome to behold.

Wang Tse's happiness knew no bounds. After some three or four days in the company of his lovely bride, however, he felt it necessary to return to his post in Peichou, and thither Blackfoot escorted him. Shortly after they had entered Wang Tse's small dwelling a knock came at the gate. On going out Wang found the adjutant of the barracks and some of the men. The adjutant had a grievance to air. 'The Prefect of Peichou is nothing but an old skinflint intent on making his own fortune,' he said. 'He has refused us pay or rations for the next month—six thousand men, and how are they expected to live?'

'Well, what are we to do about it?' asked Wang Tse.

'Revolt, if there is no other way,' was the adjutant's answer.

Wang Tse went back to tell Blackfoot what he had heard. 'Tell all the men to line up before your house here tomorrow morning,' said Blackfoot. Wang Tse did as he was bidden, and the next morning came back from the barracks to his dwelling followed by all the soldiers of the garrison, a great swarm of men. Blackfoot met him by the gate. 'Now issue them with rice yourself,' said Blackfoot.

'And where is all this rice to come from?' asked Captain Wang.

'It's in your house,' said Blackfoot: and when Wang Tse looked, there indeed was bushel upon bushel of rice overflowing every inch of every room in his house. Man after man joyfully helped himself to as much rice as he could carry away at Captain Wang's command, nor did the pile grow less until the last man was satisfied.

But that very afternoon guards came for Wang Tse and marched him before the Prefect in his court. 'What does this conduct mean, Captain Wang?' angrily demanded the Prefect. 'It is I who issue rations to the men of the garrison. Who are you, a mere captain, to attempt to usurp my authority?'

Before Wang Tse could begin to reply, attendants ran in with the news that the official granary, though the doors remained locked, had been robbed of thousands of bushels of rice. 'So that's it!' roared the Prefect. 'Stealing the Government's rice and handing it out in secret to the troops! To the cells with him, and let him be questioned under torture until we find out all about this!'

Wang Tse was taken out and flogged unmercifully. He fainted away and was revived, and, unable to bear the thought of further punishment, confessed to having received the grain now missing, from an evil magician. He gave Blackfoot's description and a warrant went out for his arrest.

Blackfoot, meanwhile, was round at the barracks haranguing the troops. He spun them a fine tale of how the wicked Prefect had himself stolen the Government's grain and sold it privately, and was now trying to throw the blame on Captain Wang. He called on them to free Wang Tse, who, he said, had so generously distributed his own fortune among them, and to rise up against the Prefect and all his minions. The men cheered Blackfoot, but none dared approach the Prefect's court to free Wang Tse. 'Leave it to me, but be ready to swear allegiance to Captain Wang as your true leader,' commanded Blackfoot. Then without waiting to be arrested he went off alone to court and gave himself up.

Now Blackfoot in turn was hauled off to the same cell where Wang Tse lay, and the guards prepared to flog him. But when the first bamboo fell on Blackfoot's back, he remained unharmed: the guard it was who felt the blow and jumped back, howling with pain! A second guard took over and received the same treatment: the blows were falling on the floggers, not the flogged! Soon all the guards had fled screaming. Now Blackfoot muttered a spell, and off came his fetters and those of Wang Tse as well. The two raced back to where the Prefect sat in state. This poor fellow, seeing them free, tried to hide behind a screen: but Wang Tse seized a sword and with one stroke cut off his head. Then he threw open the gates of the court and called in his men, and in poured like a great wave the six thousand soldiers of the garrison. They slaughtered the Prefect's guards and his underlings and every member of his household, and all the stocks of food and money and treasure in the place were seized and divided up among them.

But two minor officers of the Prefect's court succeeded in escaping the massacre. These men hastened to the Eastern Capital, and there they reported that the standard of revolt was raised in Peichou. His Majesty at once appointed a commissioner to put down the rebels and entrusted five thousand men to his charge. The man he selected was Liu Yen-wei, who had long been known for his command of strategy. When Wang Tse heard that this general was marching against him at the head of five thousand picked troops he trembled in panic, but Blackfoot and Eterna, who now had joined him in Peichou, told him not to fear, for they were at hand to assist him. As for the priest Eggborn, he had no liking for the killing that had already taken place, and would have no part in the battles to come. He retired to a small temple a little way from the city of Peichou to pass his time in quiet.

Liu Yen-wei divided his forces into three: a vanguard of one thousand men, a main army of three thousand, and a force of one thousand more to bring up supplies in the rear. Wang Tse, of his six thousand men, set half to defend the city and split the remainder into two forces, each of one thousand five hundred men, to meet the approaching enemy.

As it chanced, the commander of Liu Yen-wei's vanguard was inexperienced, and had sent out no scouts. As his small force, weary from the long march, approached within reach of Peichou it fell between the two wings of Wang Tse's attacking force and was all but wiped out. Wang Tse's men returned in triumph to their camp just outside the city wall, whilst Liu Yen-wei pitched camp some miles away with his main army and issued instructions to his second-in-command, Tuan Lei. This officer sent two small companies of foot soldiers secretly by night to take up positions near the rebel camp. The next morning Tuan Lei rode forth at the head of a further small force. Coming in view of the rebel camp outside the walls of Peichou, he drew up his men in line of battle, then rode out alone to challenge the rebels to fight. At sight of this warrior, in helmet of steel and embroidered trappings and whirling in his hands a mighty battle-axe, the rebel commanders told themselves that here was Liu Yen-wei himself. They charged out to put an end to him. They circled, thrust and parried, and soon Tuan Lei cried 'Enough!' and rode back a space. Cheering, the rebel troops followed up on the heels of Tuan Lei's men, many of whom had thrown away their weapons and fled. Catching up again with Tuan Lei the rebel commanders again tried to cut him down, and succeeded, as he fully intended, in driving him back still farther. But now as the rebels advanced again a messenger rode up with the speed of the wind. 'Pursue no farther!' this man yelled. 'Our camp is on fire!'

The night-marchers had done their work: no sooner

had the rebels been lured by Tuan Lei away from their camp than it had gone up in smoke. Furious, outwitted, the rebel commanders ordered the retreat. But even as their army milled about, there swept down on them Liu Yen-wei himself at the head of his main army. It was an irresistible wave, and more than half of the rebels' three thousand perished before it. The rest scrambled somehow or other back to the city, where Wang Tse was soon beside himself with terror.

But Blackfoot said, 'Give me five hundred men, and I will settle with Liu Yen-wei.' The clamour outside the gates announced the arrival of the Imperial forces, and Blackfoot marched out at the head of his small company to receive them.

Liu Yen-wei roared out his scorn of the short lame figure with neither mount nor armour who dared to oppose him. But exactly as he gave his order to charge, Blackfoot pointed with his little sword and cried 'Quick!': and a whirling sandstorm swept in the faces of the enemy. Blinded by flying grit and stones the Imperial troops turned away, and Blackfoot's rebels fell on their ranks and wreaked slaughter there.

Trailing back to his camp with what was left of his army, Liu Yen-wei asked himself how he was to deal with witchcraft such as this. At length he hit on a solution, and after three days' rest he led his men forth again. This time each man and each horse was equipped with a gauze mask through which not even fine sand would penetrate. They ranged themselves in front of the city wall and he challenged the rebels to come out and fight.

Again Blackfoot pointed his sword, but now it was not a sandstorm but a horde of wild beasts which swept over the Imperial forces. Leopards, tigers and wolves flew out of the air upon them, and they turned tail and fled in terror and again were pursued and reduced by the little company of rebels.

Liu Yen-wei rested his men again before one last attempt. This time, in addition to the gauze masks, he had three hundred horses hung with painted canvas to resemble lions. The lion is the king of beasts and would surely put to flight any other beast the rebel wizard loosed at them. But Blackfoot did not even deign now to carry a sword. In his hand as he led out his men from the city gate was a fan. This he waved, and cried 'Quick!': and the air froze about the Imperial troops. Unable to grip their weapons, barely able even to move, they fell by the score to the rebel attackers. And now as the cavalry on the two wings of Liu Yen-wei's army charged in, a black cloud came down so that they could see nothing. In their confusion each squadron mistook the other for the rebel force, and they hacked each other to pieces in the darkness.

Of his five thousand men Liu Yen-wei could now muster but a few hundred. Even these were weary and worn, for not even sleep at night was permitted them. All night long Eterna soared by magic about their camp, raising a threat here and an alarm there so that the sentries were constantly shouting 'Enemy attack!' and the guards ran to and fro until the day broke. Utterly dispirited, Liu knew there was nothing more he could do, and with the pitiful remnants of the Imperial force he turned for the long march back to the capital.

Within the city of Peichou all was feasting and merriment. Although he had done little enough to help himself, Wang Tse was looked on by his men as the hero of the hour, and he raised no objection when they urged him to accept the title of King of Eastern Peace and to mount the throne and rule the lands he had won. Now the prediction was fulfilled and Eterna was a royal queen. Blackfoot was made a marquis for his services, Aunt Piety was given a palace of her own as Royal Aunt, and even Eggborn the monk, who had done nothing but sit in meditation throughout, was loaded with honours in his near-by monastery. Jen, Chang and Wu now came to the city, each bringing a band of recruits. For six months the new King of Eastern Peace gave himself up to pleasure and to wickedness. The treasure of the city proved insufficient for his greed, and his men raided far and wide through the countryside to bring him anything they could lay their thieving hands on. Blackfoot, who through all his cultivation of the magic arts had never lost his taste for wine, surrendered himself to continuous orgies of drunkenness. As for Eterna, she performed none of the proper duties of a queen but amused herself by playing cruel tricks on the innocent people of her realm. Their lives became a misery as they were robbed of their livestock to furnish royal banquets, taken from their fields to build ever more luxurious additions to the palace, and cruelly tortured for the least sign of dissatisfaction with their lot.

As more and more of these things came to the knowledge of Eggborn the monk he grew increasingly unhappy. He recalled the vow that he had made in the Cave of the White Cloud. The secrets of the sacred Text of Heaven had ensured the success of Wang Tse's revolt against a miserly and unjust Prefect. But now they were being used, not for good as he had vowed, but only for the pursuit of wickedness. Surely all of them in Peichou were acting against the way of Heaven? And for this he himself must be responsible. No longer able to bear his nearness to the city of sin, Eggborn left his monastery and with a heavy heart set out to roam the earth.

If there was joy in Peichou, there was consternation in the Eastern Capital when Liu Yen-wei returned with the miserable remnants of the Imperial force. The Emperor and his ministers listened in silence to the general's woeful report. This clearly was no ordinary disturbance, but a revolt of the demons against which only a mighty effort could prevail. Accordingly, after long and careful deliberation the Emperor appointed his greatest statesman and wisest strategist, Wen Yen-po, to take personal charge of the suppression campaign. This man was of the highest virtue, the truest piety and the most profound wisdom. To him was entrusted a vast army of one hundred thousand men and all the weapons and equipment that were needed. The army moved forth in impressive array and marched to the city of Peichou.

To greet it Wang Tse opened wide the gates of the city. He stood at the head of his men close outside the wall. Blackfoot was beside him, Wu the noodle-vender on his left and Jen the pieman on his right. Chang the butcher stood on top of the city wall beating a great drum, whilst Eterna led a troop of her own guards on a tour of inspection of the defences. Wen Yen-po made his dispositions, a main body in the centre, a vanguard in front and wings to left and right, then rode forward alone to parley with Wang Tse.

'My cause is just,' shouted Wang Tse as General Wen approached. 'The Government of this region was corrupt and I overthrew it. Now I rule it in peace: why does the Emperor send troops against me?'

'Your crimes are beyond forgiveness,' returned Wen Yen-po as he drew up his mount. 'Now is your opportunity to surrender, for my army is not to be withstood.'

'A pity!' called Wang Tse now. 'You are an honourable man, and might have enjoyed a peaceful old age.'

Enraged, Wen Yen-po gave the order to beat the drums, and the great army moved forward. But as it did so, from the rebel front came black clouds and blinding rain. Thunder rolled and lightning flashed. Flying sand covered the sky, and down on the Imperial troops descended warriors with demon faces and wild beasts of every kind. Wang Tse, seeing the enemy thrown into confusion, led forward his men to attack. But as he did so the wings of Wen Yen-po's army closed in on him, and fearful of losses he retired to the city. The Imperial army beat the retreat, and when it had halted some miles from the city General Wen found that his losses numbered thousands of men either killed by the rebels or trampled to death by their own panic-stricken fellows.

He called a conference of his subordinates, and as they talked one man spoke up to give him fresh hope. 'This form of magic they are using I have myself at one time studied,' said this man. 'I am unable to practise it, but I know how it is to be countered. It is a magic which cannot withstand a certain mixture: touched by the merest spot of this mixture, it cannot act.'

'And what is this mixture you speak of?' asked the general.

'The blood of a sheep, the blood of a pig, the stale of a horse and the juice of garlic,' was the reply.

General Wen ordered the preparation of this noisome mixture to be put in hand without delay. Great buckets were filled with it, and all the spears and axes, the swords and arrows of the entire army were smeared with the concoction. Thus prepared the army advanced again, in formation as before.

Now Jen the pieman, Chang the butcher and Wu the noodle-vender complained to the rebel king that they had not yet had an occasion to show their mettle.

'Very well,' said Wang Tse. 'Today each of you shall lead a force to counter the enemy's vanguard and wings. The Marquis Blackfoot and I will lead the main body against the centre.'

Chang the butcher first led out his men against the great Imperial vanguard of five thousand. He had no weapon, but merely held in his hand his magic gourd. From this he poured forth first flood and then fire, which crashed and raged about the enemy. The vanguard could not withstand this, and turned and made off to the east with Chang and his men in pursuit. Now Wang Tse rode to face the main body of the Imperial army and Blackfoot unleashed on them his hordes of demon warriors and wild beasts. But five hundred of General Wen's picked archers lay in wait for these, and as their arrows tipped with filth found their mark the magic men and beasts disintegrated in mid-air. Wang Tse was horrified, and ordering the retreat galloped back with Blackfoot to the safety of the city.

Wu the noodle-vender had led his force to the east of the main body, where he came up against a squadron of the Imperial cavalry whose leader charged at him with lance at the ready. Wu himself in his youth had been no mean exponent of the lance, and the two fought bravely for a time. But now a junior officer came to the aid of Wu's opponent, and the former noodle-vender felt himself in peril. A pat on his horse's neck, and it rose into the air out of the ring of combat. It passed above the heads of the vanguard which Chang the butcher had routed: the archers of this force at once let fly at it with their arrows. Each arrow had been dipped in filth, and the magic horse was after all no more than a paper cut-out. Down fluttered the paper horse, down came Wu to earth with a bump, straight in the path of Chang the butcher hot in pursuit of the enemy. Chang plucked up his comrade on to his own saddle-bow, but by this time the main force of the enemy was almost on him and he turned back and bore the badly shaken Wu back within the city gates. The forces Chang and Wu had commanded, left leaderless, were soon dispatched by the Imperial swordsmen.

Jen the pieman, meanwhile, had straddled his stool, changed it into a white-browed tiger and was leading his company fearlessly to combat General Wen's other wing. The commander of this force brandished a great steel trident as his chosen weapon. At sight of him Jen the pieman patted his tiger on the head, and it leapt in the air and came bearing down snarling and with claws outstretched on the adversary. Up came the points of the trident, smeared with the potent mixture, in a prodigious thrust; and it was no fierce tiger which toppled down but an ordinary wooden stool. Jen was taken and trussed like a chicken and led off for delivery to General Wen.

The battle for the moment was at an end. The rebel forces had been decimated, and now Jen the pieman was questioned. He revealed that Eggborn, of whom General Wen had been most afraid, had left the city. He himself was now a prisoner and his two comrades were discredited. Only Blackfoot and Eterna remained, together with Aunt Piety who so far had taken no part in the fighting. General Wen now decided to attack the city itself, with cannon and battering-rams and scaling ladders, all the equipment of a siege. But the losses were too great as cloud after cloud discharged monsters and wild beasts from the city, too many for the Imperial bowmen to shoot down.

For the moment even the wise and virtuous Wen Yen-po could make no further progress against the rebels of Peichou. But now the magic storms, the floods and fire, the hordes of demons and wild animals which had been let loose about the city had raised such a stink of foulness as rose up to Heaven and reached the nostrils of the Supreme Ruler of all the spirits, the Jade Emperor himself. Disturbed and dismayed, he sent spirit-messengers to Peichou to investigate and report. From what they told him he realized that such mischief as was afoot could have come from one source only: the Cave of the White Cloud where the sacred Text was inscribed. The White Monkey Spirit, Guardian of the cave, was summoned at once. He now told the story of Eggborn, of the patience and perseverance of his threefold attempt on the cave and of his vow never to use the secrets of the Text to any end but the good of mankind.

'And now he has broken this vow,' said the Jade Emperor gravely.

'Begging Your Majesty's forgiveness,' hurriedly interposed the White Monkey, 'the monk Eggborn has had no hand in the events at Peichou. The troubles there are all to be traced to Aunt Piety and her brood of demon foxes.'

'Well, you were the Guardian of the Text,' said the Jade Emperor. 'You let it be stolen; you must now put a stop to this mischief.' And with these words the White Monkey Spirit was dismissed from the ruler's presence. His first thought was to seek out the monk Eggborn. He found him in holy seclusion on the highest peak of the Red Gold Mountain, far from any living thing. Eggborn was shamefaced to recall the last meeting of the two of them, when the White Monkey himself had assisted him in his quest for the Text. He showed himself only too eager to give what help he could now in suppressing the revolt of the demons. To encourage him, the White Monkey Spirit showed Eggborn a precious aid which he had begged from the Jade Emperor himself. This was the celestial mirror of exorcism, which shown to a demon in any disguise would strip the disguise away and reveal the true form underneath. Eggborn expressed his gratitude for the trust which had been placed in him, and returned to Peichou.

There Eterna was proposing to Wang Tse a plan to be rid of the Imperial General Wen Yen-po, who with his bombardment and his fire-arrows was making life almost impossible within the beleaguered city. She had men find the biggest millstone in the city, a huge block of stone a foot or more thick. On this with a brush she wrote magic directions. She spoke a spell, spurted water on the millstone and cried 'Quick!'

General Wen at this moment was seated outside his tent discussing plans with his subordinates. Suddenly a mighty rushing wind disturbed them. Looking up they saw a huge millstone flying through the air towards them. It came faster than thought, straight for the chair on which the general was sitting. But even as it fell, a strong arm plucked the general from his chair and drew him aside. A second later the chair was in splinters and buried feet deep below the gigantic stone.

'Who are you?' asked Wen Yen-po of the man who had saved his life: for it was none of his army. The man asked for a brush and wrote down his name. He covered the paper with an inverted bowl and strode away. After a pace or two he vanished. 'Another monster!' cried the officers; but General Wen had by now read the name on the paper. The name was Hundred-Eyes, and the general was puzzled for some time; but then it came back to him, how once years ago he had dreamed of a ghostly visitor. This figure had announced itself in his dream as the spirit Hundred-Eyes. The spirit was in disgrace in Heaven for a minor fault, and needed prayers for its forgiveness. Ever a generous man, Wen Yen-po on waking had prayed in the nearest temple for forgiveness for Hundred-Eyes, who had appeared again the next night in a dream to thank Wen for his mercy and to promise assistance in the future. And now here was the promised aid. General Wen told all his staff the strange story, and was filled with joy to know that the spirits of Heaven were now fighting on his side.

Eterna, her husband and her brother turned now in

their dismay to Aunt Piety for help. Aunt Piety ordered the King of Eastern Peace to lead out his men once more. At their head, as they faced the Imperial army outside the gates of Peichou, stood Wang Tse himself, Blackfoot and Eterna, Chang the butcher and Wu the noodle-vender, each one with his eyes fixed on the one old woman to see what new magic she would now perform. With beating drums and clashing cymbals Wen Yen-po's army rode to the attack. Unhurriedly Aunt Piety led forward a white horse. With the point of a sword she pricked its skin to draw blood, which she let fall into a bowl of water. The mixed blood and water she sprinkled on the ground, then uttered a spell and cried 'Quick!' At once the ground shook and split, the sky went black and thunderbolts rained down upon the enemy.

The Imperial ranks were thrown into chaos. General Wen himself could only manage to control his own mount: he could see nothing, hear nothing, do nothing but keep from falling in the tumult all about him. Then suddenly a whirlwind sprang up before his horse's head and a bright light appeared before him. The whirlwind moved forward and he followed it, riding down a path of light. For miles he rode, away from the field of battle, compelled to follow the light. It led him away from the city, to a little valley where finally his horse came to a halt at the gate of a small, neatly kept monastery. General Wen dismounted and entered the gate. An acolyte led him into the presence of the holy abbot, who was none other than Eggborn. The abbot greeted the general by name, bade him be seated and gave him tea to drink.

'How is it that you know my name?' asked General Wen.

'I have long known of your great virtue,' replied

Eggborn, 'and I have been awaiting your arrival here. But tell me, why is it that with your profound knowledge of strategy and your great force of a hundred thousand men you have not yet succeeded in putting down this miserable rebellion at Peichou?'

Wen Yen-po recited all the details of the magic tricks the rebels had used against him. 'I know what kind of magic arts are these,' announced Eggborn when he had finished, 'and I think I can do something to help you. Come, rest here for a while, and then I will ride back with you to your camp.'

They found the Imperial army counting its losses, which were the heaviest yet suffered.

Next day Wang Tse again led his men out of the gates, determined this time to finish off, with Aunt Piety's help, what remained of General Wen's forces. As before, Wang Tse and the rest of his assistants did nothing whatever but stand beneath their banners and watch as Aunt Piety led out her white horse, drew its blood and performed her incantations. The great storm crashed and rolled; but now a most weird change took place. At the head of the Imperial ranks came Eggborn, gently ringing his handbell: and the storm stopped in its course, then, without ever abating, reversed and thundered back towards the massed rebels themselves! Real warriors and real weapons, demon warriors and demon weapons, all alike were picked up and tossed back against the city. It was all Wang Tse could do to regain safety inside the city wall with such of his followers as were not struck down by thunderbolts or by terror or by the Imperial troops who now pressed forward in hot pursuit. Somehow the gates were shut and the walls manned to repel attacks from the besieging army.

Still there was no way of gaining entrance to the city.

General Wen was profuse in his thanks to Eggborn for his invaluable help and discussed with him what was to be done now. As they were speaking there came forward an officer of Wen Yen-po's army, a certain Captain Ma. He, it proved, was a fellow-townsman and old acquaintance of Wang Tse the rebel leader, whose conduct he abhorred: and now he asked leave to be sent as a messenger to the rebel city, where he declared he would gain the presence of Wang Tse and assassinate him. Deprived of its leader, the rebellion could not then last long.

Such a courageous plan at once won the assent of General Wen, and the next morning Captain Ma rode alone to the city gate. As the bearer of a message he was admitted before Wang Tse, who, although the message itself was merely a fresh challenge to battle, suspected nothing of Captain Ma's true intent. He greeted him as an old friend, and accepted his story that the Imperial army was now much enfeebled and that he himself had decided to throw in his lot with the rebels. But try as he might, Captain Ma could never get Wang Tse alone, away from the company of his entourage of magicians. They talked and feasted, and that night Captain Ma stayed in the room of Chang the butcher, with whom he sat up drinking till the small hours.

Now Chang the butcher was impatient to make further use of his magic gourd, and when the next

Imperial attack came he pleaded to be allowed to meet it. With him he took Wu the noodle-vender and no one else, for Blackfoot was drunk at the time and Eterna and Aunt Piety both resting in their palaces. Chang walked through the city gate to face the enemy, opened his gourd and cried 'Quick!': and nothing at all happened. 'Quick, quick!' he cried; but neither flood nor fire appeared. For this was the doing of Captain Ma, who during the night, while Chang slept, had filled his gourd with the mixture of blood and filth and garlic which he carried in a tiny flask inside his girdle.

Wang Tse, though married to Eterna, had only ever succeeded in learning one spell, which was to stop an enemy in his tracks so that he froze and was unable to move hand or foot for an hour. Now, watching from the city wall, he saw General Wen's forces bear down on the defenceless Chang the butcher and on Wu the noodle-vender, frantically looking for a stool to change into a magic tiger. The King of Eastern Peace opened his mouth to utter his spell. But this was Captain Ma's moment, and he flung himself upon the rebel leader. The brave captain was weaponless, and the royal bodyguard were on him at once; but not before he had dealt Wang Tse such a blow as knocked out all his front teeth. The gallant captain had failed in his assassination attempt and was executed on the spot; but he had rendered Wang Tse powerless to speak any spell, and now beyond the wall Chang and Wu and all their force were slaughtered to the last man.

Aunt Piety was summoned in haste to the bed where the King lay recovering from his blow. Her expression was grave. 'There is only one thing for it now,' she said. 'I must go for the Ultimate Magic, a thing only to be used in time of the greatest peril.'

'What is this thing?' asked Wang Tse, mumbling through his broken teeth.

'It is a black sword which I made years ago when I was studying with my son and that accursed monk Eggborn,' said Aunt Piety. 'Its use is to behead a black dog. Then, when the dog's head falls, so does the head of any man whose name has been pronounced beforehand. This is the way to finish with Wen Yen-po, and with all his officers and with all his men too if need be.'

'And where is the sword?' asked Wang Tse again.

'It is where I hid it away, beneath a rock on the highest peak of the Pillar of Heaven Mountain. I shall go now and fetch it, and when I return we shall see its terrible effect.' And with these words she left.

Meanwhile a second brave officer of the Imperial army had come forward with a stratagem. This was to lead a company of five hundred engineers to dig a tunnel beneath the city wall, all attempts to scale which had failed. The exact lay-out of the city had been learnt from the captured Jen the pieman, and with careful direction the tunnellers should come up within the royal palace itself. The monk Eggborn offered to accompany the tunnellers in order that he might counter any magic which might be used against them when they emerged.

Transporting herself by magic means, Aunt Piety soon reached the peak of the Pillar of Heaven. But there to her astonishment she found, sitting on the very rock beneath which the Black Sword lay, an old man with a long white beard: it was the White Monkey Spirit, who had foreseen this ultimate move of Aunt Piety's. She did not know him, and greeted him peaceably.

'What do you seek here, old lady?' asked the White Monkey.

'Only a sword that once I left here,' she replied, and

bending down took out the Black Sword from its place of concealment.

'Let me see it,' said the old man. Aunt Piety handed him the sword, inwardly cursing in her impatience to be off. He stroked his hand along the blade. Then, 'It doesn't seem to be in very good condition,' he said, and passed it back to her. It was nothing now but an ordinary old sword, half rusted away from disuse. Aunt Piety knew she had been tricked. Furious, she flew at the old man. But from under his robe he drew out an embroidered bag. When Aunt Piety saw it she knew what it contained: the celestial mirror of exorcism. Any demon or monster of whatever kind, catching sight of its reflection in this mirror, would change at once, she knew, back to its proper form. Before the White Monkey was able to draw out the mirror from its bag she had shut her eyes tight. Yet, though she was able to retain her form as a venerable old lady, she was rendered powerless by the mirror. The White Monkey Spirit called for heavenly guards, who dragged off Aunt Piety before the throne of the Jade Emperor for judgement.

Beneath the city of Peichou the tunnellers now reported the completion of their task. Their captain, with Eggborn at his side, was the first to emerge. He found himself standing inside the palace, in the passage outside the bedroom of Wang Tse himself. Into the room he burst with his men. The royal bodyguard cowered in terror, and the King of Eastern Peace, still unable for lack of his teeth to mumble his only spell, was seized and tied hand and foot for delivery to the Eastern Capital.

From the royal bedchamber the invaders now sought out that of the Queen, Eterna. As they approached they found it guarded by a magic sea, which dashed up spray in their faces and seemed impassable. But Eggborn spoke a holy prayer and the waves disappeared. Into the room they dashed, and seized and bound Eterna, robbed of her powers by the filth they had first sprinkled over her. Dragging her out they set fire to the entire rebel palace. Then they went out into the streets of the city, where the townspeople of Peichou, liberated at last from the long oppression of the demons, assisted them in tracking down the unrepentant rebel soldiers. General Wen Yen-po knew from the flames that rose to the sky that the tunnellers had done their work in the palace. He led in his army through the now wide-open gate, and with his great army made short work of restoring order in the city and issuing food to all the victims of the siege.

There still remained Blackfoot to account for. He was nowhere to be seen. Puzzled, Eggborn made his way to the palace which had been built for the Royal Aunt, Aunt Piety. The palace was deserted. Standing in the great hall, Eggborn let his gaze travel slowly round the four corners. Standing against the wall in the farthest corner of the hall was a curious object to find in such a place: it was an old broom. Eggborn stared at it. An ancient man with a long white beard, the White Monkey Spirit, had now appeared by Eggborn's side, and the monk turned to him with a questioning glance.

'Yes, that is Blackfoot,' said the White Monkey. 'Let us see him as he really is.' And he took out the celestial mirror of exorcism and held it to face the broomstick. At once the broomstick disappeared, and in its place a red fox with one lame foot cringed against the wall. Only for a second: then a thunderbolt fell from the air, and Blackfoot the fox lay dead. A cry of wonder arose from outside; and going out to see what had happened, Eggborn found that Eterna, at precisely the same moment, had likewise been struck by a thunderbolt. In place of the captive queen lay the red fox-carcass of Blackfoot's sister.

General Wen Yen-po now sought out Eggborn to express on the Emperor's behalf the nation's gratitude. But Eggborn had gone. In company with his old friend the White Monkey Spirit he had left the mortal scene to make his report to the Jade Emperor, the ruler of all the spirits.

The revolt was over. General Wen led his army and his captives in triumph back to the Eastern Capital, where Wang Tse was committed to a small cell to await slow and painful execution. Peace reigned once more in the city of Peichou, and the farmers of the surrounding countryside could once again till their fields undisturbed.

And Aunt Piety? Her true undoing was her meeting with the White Monkey Spirit on the peak of the Pillar of Heaven. For now the Jade Emperor decreed that the White Monkey had served his full term as Guardian of the Cave of the White Cloud. He was now to return to the company of Heaven, and his place was to be taken by—Aunt Piety.

The old sorceress felt herself lucky to have escaped with so lenient a sentence for all her misdeeds. But she was not to know. No sooner had she reached her new post in the cave in the folded, mist-shrouded hills than the earth shook and the mountains heaved. A huge landslide blocked the mouth of the cave in which Aunt Piety sat and sealed it up for ever. And nothing more was ever heard again of her or of the Text which she so faithfully guarded.